X-PLANES 5

TSR2
BRITAIN'S LOST COLD WAR STRIKE JET

Andrew Brookes

SERIES EDITOR TONY HOLMES

OSPREY
PUBLISHING

OSPREY PUBLISHING
Bloomsbury Publishing Plc

Kemp House, Chawley Park, Cumnor Hill, Oxford OX2 9PH, UK
Bloomsbury Publishing Ireland Limited,
29 Earlsfort Terrace, Dublin 2, D02 AY28, Ireland
1385 Broadway, 5th Floor, New York, NY 10018, USA
Email: info@ospreypublishing.com
www.ospreypublishing.com

OSPREY is a trademark of Osprey Publishing Ltd

First published in Great Britain in 2017
Transferred to digital print in 2024

© Osprey Publishing Ltd, 2017

All rights reserved. No part of this publication may be: i) reproduced or transmitted in any form, electronic or mechanical, including photocopying, recording or by means of any information storage or retrieval system without prior permission in writing from the publishers; or ii) used or reproduced in any way for the training, development or operation of artificial intelligence (AI) technologies, including generative AI technologies. The rights holders expressly reserve this publication from the text and data mining exception as per Article 4(3) of the Digital Single Market Directive (EU) 2019/790.

A catalogue record for this book is available from the British Library.

Print ISBN: 978 1 4728 2248 2
ePub: 978 1 4728 2249 9
ePDF: 978 1 4728 2247 5
XML: 978 1 4728 2250 5

Edited by Tony Holmes
Cover artwork, Battlescenes and Aircraft Profiles by Adam Tooby
Index by Mark Swift
Page layouts by PDQ Digital Media Solutions, Bungay, UK
Printed and bound in Great Britain by CPI (Group) UK Ltd, Croydon CR0 4YY

MIX
Paper | Supporting responsible forestry
FSC® C013604

25 26 27 28 29   12 11 10 9 8 7 6

Product safety
For product safety related questions contact productsafety@bloomsbury.com

The Woodland Trust
Osprey Publishing supports the Woodland Trust, the UK's leading woodland conservation charity.

www.ospreypublishing.com
To find out more about our authors and books visit our website. Here you will find extracts, author interviews, details of forthcoming events and the option to sign-up for our newsletter.

Acknowledgements
I am very grateful to everyone who has trawled back to recall events that took place more than half a century ago. In particular, I have leant very greatly on the incomparable Jeff Jefford of the RAF Historical Society, Andrew Lewis of the Brooklands Museum and Dave Ward of the Warton Heritage Centre. Many thanks to you all – this book could not have been written without you.

Front Cover:
XR225 trials and test aircraft in white reaching Mach 2 at 30,000ft over the Irish Sea. Note that aircraft painted anti-flash white had roundels and fin flashes in pale colours.
(Artwork by Adam Tooby)

Title page
XR219 in flight.
(BAe Systems Heritage)

XPLANES
CONTENTS

CHAPTER ONE
INTRODUCTION 4

CHAPTER TWO
ORIGINS 5

CHAPTER THREE
THE AIRFRAME 10

CHAPTER FOUR
ENGINES AND AVIONICS 18

CHAPTER FIVE
FLIGHT TESTING 38

CHAPTER SIX
THE POLITICS 52

CHAPTER SEVEN
WHAT MIGHT HAVE BEEN 66

FURTHER READING 79
INDEX 80

CHAPTER ONE

INTRODUCTION

When I learned to fly with Leeds University Air Squadron, my aim was to fly the TSR2. I still recall the sense of shock when it was cancelled. The Royal Air Force (RAF) and the British aeronautical industry have suffered many aircraft and equipment project cancellations since 1945, but none caused as much anguish and debate as the cancellation of TSR2. Marshal of the RAF Sir Michael Beetham was in the Operational Requirements branch of the Air Ministry in the early 1950s when the Canberra was getting well established in service and the V-bombers were coming in. 'As a young squadron leader, I was told by my boss to pick up my pencil and to write the first draft of what became OR339, for a replacement for these aircraft, especially for the V-bombers in the strike role. I bit the end of my pencil and did my best, and in due course I left that job and passed on to other appointments. I still feel, however, that the cancellation of TSR2 left a gap that was never satisfactorily filled.'

More than 50 years later, TSR2 still raises blood pressures to a much greater extent than the cancellation of the Javelin Mk 3 or the HS681 Short Take-Off and Landing (STOL) military transport. So I will examine the background and history to determine whether TSR2 was a white elephant or a potential world-beater. Did the RAF over-egg the specification and was the cancellation 'political'? Where did the fault lie for 'failing' to proceed with the project, if fault there was? What did TSR2 leave behind in terms of research and development, systems and military thinking? And if it had entered service, what difference might TSR2 have made to operational history?

TSR2 under the spotlight. (Brooklands Museum)

CHAPTER TWO

ORIGINS

The previous generation. A Canberra B(I)8 and USAF F-105 Thunderchiefs from Spangdahlem, in West Germany, fly in formation over the No. 16 Sqn Operational Readiness Platform sheds at Laarbruch, also in West Germany, in the summer of 1964. The B(I)8 pilot is Mike Kelly and the photograph was taken by Graham Pitchfork from a No. 31 Sqn Canberra PR7. (Graham Pitchfork)

The first formal paper proposing an Air Staff Requirement for a Canberra replacement circulated within the Air Ministry in November 1956. It called for a tactical strike/reconnaissance aircraft capable of supporting a tactical offensive (possibly nuclear) in limited or global war. The intention was to exploit a combination of high speed with low altitude to ensure that all possible advantage would be gained from the difficulties which an enemy would face in erecting an effective defence down low. The Air Staff envisaged that the aircraft would also be capable of carrying out alternative medium and high altitude strike and reconnaissance missions when the air situation was favourable. Thus, the primary role of the aircraft was to deliver tactical nuclear and high-explosive weapons from low altitude up to the maximum radii of action obtainable, in all weather conditions by day and night. It was also to be capable of low and medium level tactical photographic and radar reconnaissance by day and night.

This outline proposal was accepted by the Air Council, and General Operational Requirement (GOR) 339 was drafted over the next few months. Issued in March 1957, GOR339 saw the basic size of the aircraft, with its two-man crew (pilot and navigator), determined by the required radius of action. Based on possible targets in Europe and obligations under the Baghdad Pact and the Southeast Asia Treaty Organisation (SEATO), the Air Staff stated a requirement for a radius of action of 1,000 nautical miles (nm), of which 200nm were to be at low level.

Main base vulnerability from surface-to-surface Soviet Intermediate Range Ballistic Missile (IRBM) attack led to a requirement for dispersed operations from short runway airfields. A takeoff ground roll of 1,000 yards was stipulated (there were said to be 1,000 airfields of this size in Western Europe), though this was reduced to 600 yards for a short-range mission. Invulnerability was also required in the air. To enable medium level strike and reconnaissance missions to be flown in airspace defended by supersonic fighters, a requirement was laid down for sustained flight at supersonic speeds.

A fully automatic navigation system had to provide an accuracy of 0.3 per cent of distance gone from the last fix. Sideways-looking radar (SLR) was to update the computer automatically, which, with a fix 30nm from the target, could give a nuclear lay-down accuracy of better than 600ft 50 per cent circular area probability (CEP). Automatic terrain following was a prime requisite, plus a low gust response for minimum crew fatigue, a long structure life and a high thrust to weight ratio. Perhaps the most crucial line in GOR339 was the seemingly innocuous statement that 'no single failure should prevent the crew from bringing the aircraft safely back to base'.

The GOR was an extremely complex and demanding requirement. The RAF put everything into the pot because, in the words of Air Vice-Marshal Wallace Kyle, Assistant Chief of the Air Staff (Operational Requirements), 'this replacement aircraft must have the greatest possible flexibility'. The introductory note to GOR339 when it was circulated added a new strategic dimension to its proposed capabilities. 'Such an aircraft', said the last of six paragraphs summarising Air Staff thinking on the project, 'with range increased by flight refuelling, would pose a low-level threat to Russia and thus augment the primary deterrent.' However, no word about strategic deterrent capability occurred in the operational requirement itself.

Nine companies responded with design proposals to GOR339 after it was sent out in the autumn of 1957 – A V Roe and Co (which submitted a heavy brochure describing the Avro 739); Blackburn and General Aircraft (which proposed a supersonic development of the NA39 Naval strike aircraft, to become the Buccaneer); Bristol Aircraft (which submitted the Bristol Type 204); Fairey Aviation (which was sent GOR339 for study, although no record exists of its proposal); English Electric (which made a thorough study of the requirement and probably expended more effort than any other firm engaged on GOR339); Handley Page (which produced a detailed study but said that it was not proposing to tender); de Havilland (which proposed the P1129), Hawker Siddeley and Vickers (which sent in a brochure). These, together with research establishment reports on instrument and electronic systems, were studied by a joint Ministry of Supply/Air Ministry Assessment Group, and in June 1958 a draft Operational Requirement was approved that detailed the aircraft's characteristics and set a Release to Service in 1965.

After much discussion, Vickers and English Electric were chosen in January 1958 as joint main contractors. Considering that they had neither submitted a joint design proposal nor even similar designs, this might seem bizarre. In fact the stimulus was purely political. In December 1957 the Minister of Supply (MoS) told parliament that 'the power of awarding contracts should (be used) to bring about a greater degree of integration'. Amalgamation was to be a prerequisite for the award of what was to be the TSR2 contract, and the choice of design was subordinated to the rationalisation (and slimming down) of the aircraft industry.

By 1959 the contractors (soon to become the British Aircraft Corporation), the Air Ministry and the MoS had agreed on what was required and what was feasible. Consequently, GOR339 was withdrawn and replaced by ASR343. Much stayed the same, but it appeared that the Air Staff had been persuaded that they had not set high enough standards. The most important of the changes were that the low-level height was redefined as 200ft or lower, speed at 40,000ft was Mach 2.0 instead of Mach 1.7 (this was described as a cheap bonus as the thrust was available due to the takeoff requirements), Radio Countermeasures (RCM) were added, ferry range was increased to 2,500nm (another supposed free bonus from the basic operational sortie requirement) and the Load Classification Number (surface load-bearing index) was reduced from 40 to 20 – i.e. from concrete to firm grass to meet the short takeoff dispersal philosophy. Yet not even Hawker Siddeley proposed using vectored thrust to mitigate the problems posed by the short field requirement.

The first public acknowledgement of the existence of the project was made on 17 December 1958 in a written answer by the Secretary of State for Air to a question raised in the House of Commons. It simply stated that the Government had decided 'to develop a new strike/reconnaissance aircraft as a Canberra replacement. This will be capable of operating from small airfields with rudimentary surfaces and have a very high performance at all levels.' This was the first time in public that the aircraft was referred to as TSR2, which, it was said, denoted its roles (tactical strike/reconnaissance) and speed of Mach 2.

A further amplification was made on 1 January 1959 by the Minister of Supply. He stated that, subject to satisfactory negotiations, the new aircraft 'would be undertaken jointly by Vickers-Armstrongs and English Electric [EE], the main contract being placed with Vickers-Armstrongs and work being shared between the two companies on a 50–50 basis. A joint project team drawn from both companies is being established at Vickers' works at Weybridge for the execution of the project.' The report continued, 'Subject equally to satisfactory negotiations, the development of the engine for the new aircraft will be undertaken by Bristol Siddeley Engines, the new company formed out of Bristol Aero Engines and Armstrong Siddeley Motors. These engine companies have indicated that they are now proceeding to a complete financial integration.

'The TSR2 is a tactical support [sic] and reconnaissance aircraft. In the course of study it has been found technically possible to incorporate in the final operational requirement modifications which will greatly increase the usefulness of the aircraft in limited operations and for close support of the Army... While the TSR2 will be capable of performing the roles of all the various marks of Canberra, it will by reason of its greater flexibility and higher general performance be far more versatile and more in the nature of a general-purpose tactical aircraft.'

The Air Staff had originally expected that the project would be allocated a number similar to that given to the V-bombers (B35/46 – e.g. B/58), but TSR2 was the designation by which it became universally known. Its operational requirement number, however, changed, the Air Staff deciding in February 1959 to call it OR343 – this Requirement, for a 'tactical strike and reconnaissance weapon system', was issued on 8 May 1959. The aircraft required in OR343, which had 114 paragraphs, compared with 48 in GOR339, was to be twin-engined. It was also to have a radius of action of at least 1,000nm without in-flight refuelling or overload fuel tanks, be able to obtain reconnaissance information for tactical purposes by day and by night under all weather conditions, deliver tactical nuclear weapons from low altitude or high-explosive weapons as an alternative, be able to operate in any part of the world and to be flown by a two-man crew.

The Air Staff required official release for the weapon system to enable a squadron to be fully equipped by the end of 1965. To complicate matters, internal equipment had to include a new high-resolution radar, active Linescan and a full photographic reconnaissance suite. The Air Staff insisted on first priority for the strike version, while agreeing reluctantly that release in the reconnaissance role could be given at a later date.

In short, OR343 called for an aeroplane that would do virtually everything in all weathers from long range nuclear deterrent strike, through all phases of reconnaissance, down to battlefield support, with a performance in excess of the most advanced jet interceptor (the Lightning) then on order for Fighter Command. Not surprisingly the writing was already on the wall for those able to decipher it. In October 1959, the retiring Controller Aircraft (CA) advised Chief of the Air Staff (CAS) Marshal of the RAF Sir Dermot Boyle that development cost estimates for TSR2 were likely to amount to 'something between £70–80m' – about twice the original estimate made in 1958.

R is for Reconnaissance

A TSR2 in the reconnaissance role, fitted with the 'reconnaissance pack'. Although the TSR2 had a permanently fitted vertical camera and oblique camera in the forward fuselage, as well as a sideways-looking radar for navigation, the 'recce pack' that could be fitted into the bomb-bay added specialist cameras, radar and Linescan for reconnaissance missions. This aircraft belongs to No. 13 Sqn.

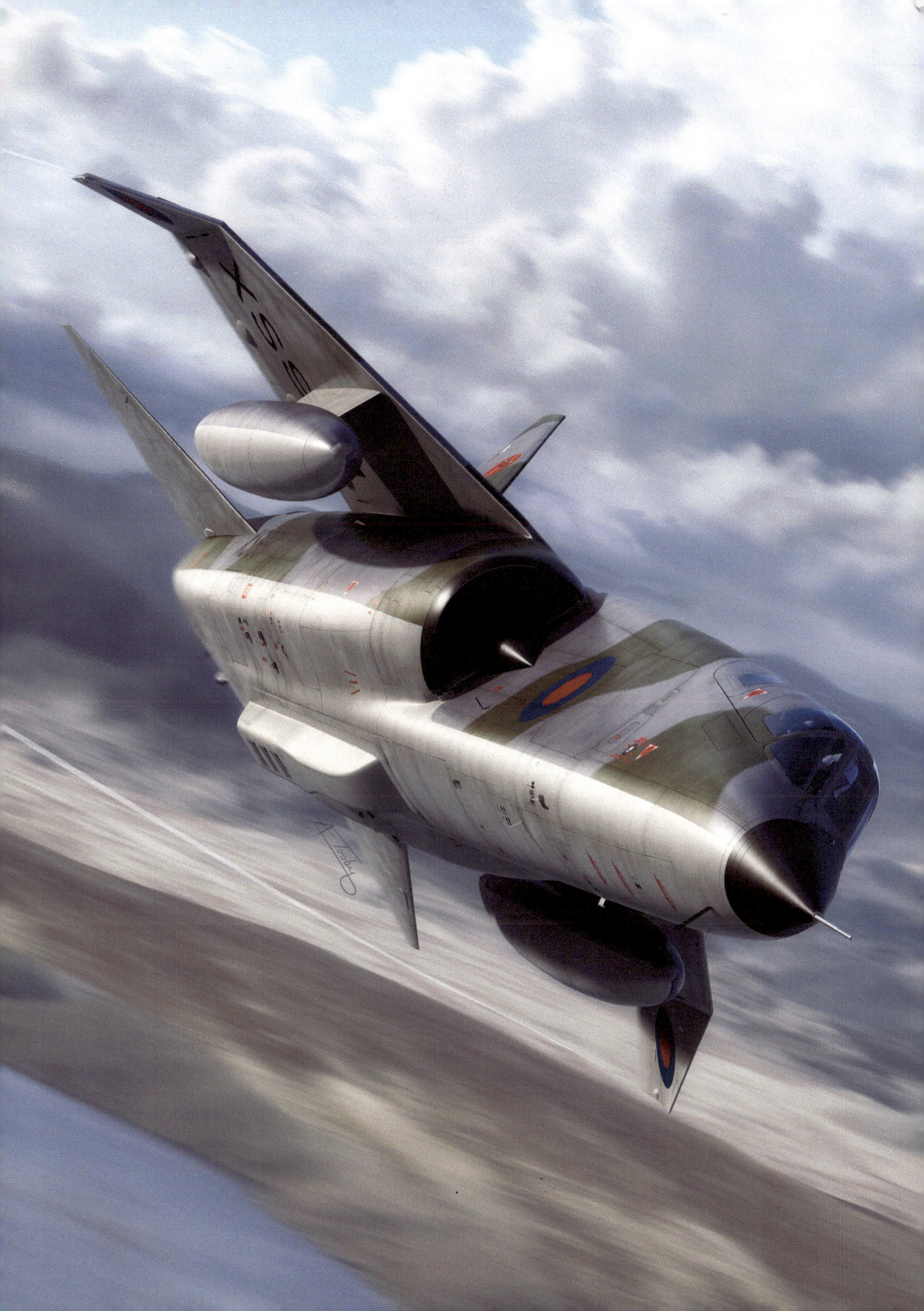

CHAPTER THREE

THE AIRFRAME

Come the 31 January 1958 deadline for submissions to meet GOR339, EE at Warton, in Lancashire, led the field because of its Lightning design experience and the amount of private research it had done into what the company termed 'Possibilities of a Multi-Purpose Canberra Replacement – the P17'.

One of the basic GOR339 requirements was for an aircraft capable of attacking at low altitude and at high speed. Under these operating conditions, the aircraft and crew would be more subject to turbulent air or gusts than at any other point in the aircraft's flight envelope. The effects of turbulence result in a shortening of airframe fatigue life, a reduction in crew efficiency and deterioration in the aircraft's handling qualities. Therefore, the need to provide a controllable aircraft that would have a minimum response to gusts so that the crew could work in comfort to navigate and deliver their weapons accurately became a crucial element of any design to fulfil the RAF's requirements.

In the 1950s there was little information available on this operational aspect, so reliance had to be placed on extrapolations of performance data from a number of aircraft that had been flown in low-altitude turbulence. These comparisons enabled EE to judge a particular configuration's response to gusts and to determine a maximum vertical acceleration, which the crew could comfortably endure in turbulent conditions. A moderately high wing-loading was considered to be the optimum, so a significant contribution to crew comfort was the reduction in lift curve slope, achieved by choosing a wing with the lowest practicable aspect ratio. An aspect ratio of about two was considered to be the optimum.

The English Electric P17A design from around 1958. The P17A wing area was 610sq ft and the maximum all-up-weight was 73,400lb. (BAe Systems Heritage)

Two configurations based on sketches drawn on 19 October 1956 by Ollie Heath (then chief project engineer) and considered to represent extremes in conventional construction were investigated. The first sketch was of an aircraft with a shoulder-mounted straight equi-tapered wing, two podded engines suspended below the wing and a long slender fuselage seating the crew in tandem behind a nose-mounted radar and with a bomb-bay and fuel tanks arranged amidships like those of the Canberra. The rear of the fuselage was left free for equipment and, in particular, for the scanning aerials required for electronic reconnaissance. The tail surfaces were conventional, with the tailplane mounted mid-way up the fin. A bicycle undercarriage stowed in the fuselage with outrigger wheels carried in the engine nacelles was envisaged.

The second sketch showed an aircraft of canard layout. The two engines in this design were positioned close together and side by side in the extreme rear of the fuselage, below the shoulder-mounted wing, and had their intakes located on either side of the fuselage and slightly behind the wing leading-edge. A straight wing planform was adopted for this aircraft and it had highly swept foreplanes.

Initial wind tunnel tests were made on what EE designated the P17, with a delta wing having a leading-edge sweep of 60 degrees and a single-slotted flap. The results obtained dictated that the engines be installed in the fuselage to decrease drag and improve lift and a low-set tailplane be fitted to achieve satisfactory longitudinal stability. Continued testing of the delta wing revealed that proper shaping of a highly swept leading edge could give unusually low drag, compared with other thin low-aspect ratio wings, both for subsonic cruising and supersonic speeds. The canard layout on the other hand was found to require a centre of gravity so far forward of the wing aerodynamic centre that the foreplanes could not trim the aircraft satisfactorily when the flaps were up, and not at all when the flaps were down. This led to the decision to opt for a delta wing.

Thus, by January 1957, the EE design team under Chief Engineer Freddie Page had come up with a Canberra replacement with a delta wing, two seats in tandem, twin RB133 engines with reheat incorporated side by side in the rear of the fuselage and with their intakes below the wings, and a low-set tailplane. To denote the departure from podded engines the aircraft was redesignated P17A. This was an admirable attempt to fulfill the requirements of OR343 in that the aeroplane was based on a sensible multi-purpose airframe using equipment in existence or under development – more advanced electronic items could be introduced later as they were perfected. P17A was to bear a strong resemblance to the TSR2 that was eventually built.

GOR339 was redrafted and re-issued no fewer than four times before it was renumbered and published as OR343. Broadly, the changes most affecting the aircraft's design during this period were increase in penetration speed and speed at altitude; a shortening of takeoff distances demanding more thrust; the ability to land in higher cross-winds, requiring more control during the approach and landing; operations from less hard semi-prepared surfaces, necessitating larger diameter low-pressure undercarriage tyres; an increase in ferry range, requiring extra fuel; and a decrease in the altitude at which the aircraft was to penetrate enemy defences, leading to more advanced and highly reliable flying control systems.

These new requirements imposed severe demands on airframe and equipment. For example, flight at sustained supersonic speeds both at low and high altitudes necessitated economic fuel consumption at high thrust, the use of materials able to withstand the strength-reducing and creep-inducing effects of kinetic heating at these speeds, and pin-point navigational and bombing accuracy which required complex equipment including miniaturised digital computers. As a result of all of these requirements, the aircraft grew in size and weight.

Meanwhile, coming in from left field was the Vickers-Armstrongs team largely composed of former Supermarine personnel brought across to Weybridge, in Surrey. Led by George Henson, it had proposed a single-engined twin-seat design using 'blown flaps' first developed on the Supermarine Scimitar and later adopted for the Blackburn Buccaneer. In effect this was a supersonic Buccaneer, and Supermarine proposed a twin Rolls-Royce-engined version, the Type 571, which was built around the GOR339 requirement for an integrated terrain-following, navigation attack system able to counter the next generation of Soviet ground radar and missile systems. Vickers (Supermarine) was the only design team to offer a fully integrated airframe-engine-avionics-weapons system.

Initial technical discussions at Weybridge in November 1958 compared the merits of the two companies' projects to determine how to merge the best features of each to produce a common design. EE evaluated all the aerodynamic data of the two designs while Vickers examined the overall weapons system. Because the general features of the individual projects were similar, it was agreed that the P17A's delta wing, together with its blown flaps to meet short takeoff requirements, should be grafted onto the Vickers design. By 1 January 1959, the firms were ready to proceed with the design of the TSR2.

BUILDING THE BEAST

A full development contract for the first nine aircraft was received in October 1960. The lateness of this contract and financial cover for materials was to presage a number of delays which accumulated and resulted in the TSR2's first flight taking place 18 months behind schedule. Meanwhile, the formation of the British Aircraft Corporation (BAC) was announced on 1 January 1960. Vickers became responsible for the nose, centre fuselage and the avionics and systems incorporated in the forward portion – to be built and assembled at Weybridge. EE was to design and produce the wing, rear fuselage and empennage, all fully equipped, and be responsible for the powerplant installation.

The rearmost fairing was to be made by Bristol Siddeley and the Vickers-designed undercarriage was produced by Electro-Hydraulics Ltd. Responsibility for design integrity was shared, Vickers taking the cockpits, overall structure, avionics and electrical and air systems, and EE taking aerodynamics and the hydraulic and fuel systems.

Structurally, the TSR2 consisted of five main elements – two centre fuselage sections, nose section housing the two crew in tandem, tail section with fin and tailplane and the complete wing. The light alloy fuselage structure consisted mainly of skin stringer panels supported by transverse bulkhead frames. Wide use was made of machined panels and chemically etched skins.

Since some 80 per cent of the total fuel load was carried in the fuselage, the structure was designed as an integral tank, with fuel in the forward tanks extending aft from the equipment bay to a point between the air intakes, and with rear tanks built round the engine tunnels. The nose fuselage contained the forward-looking terrain-following and sideways-scanning navigation radars, with the pilot and navigator in separate tandem cockpits. The main avionic equipment

The TSR2 structure comprised a forebody, forward centre-section, rear centre-section, rear fuselage, wing, tailplanes and fin, undercarriage, braking parachute and airbrakes. (Brooklands Museum)

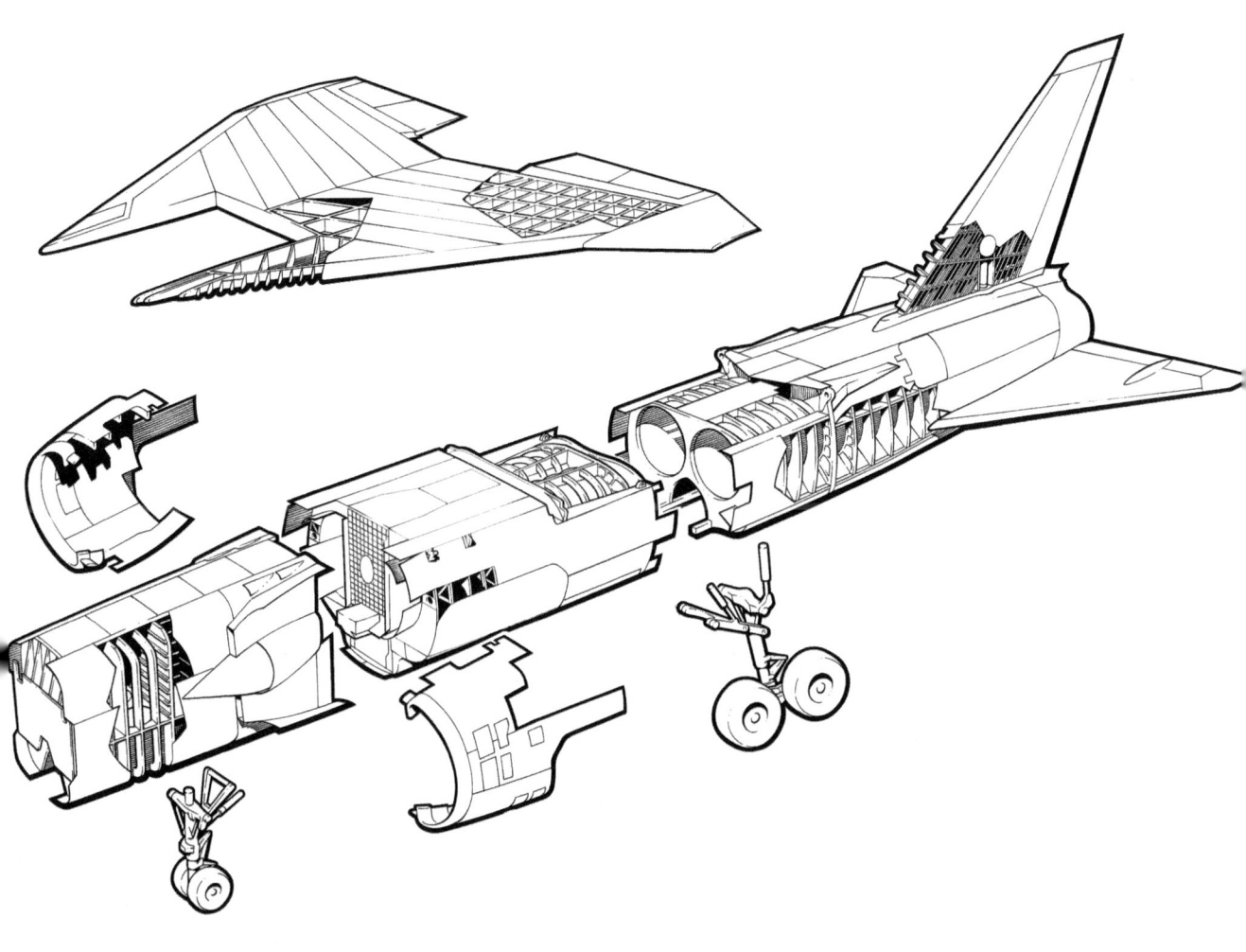

CHAPTER THREE THE AIRFRAME

The TSR2 rear fuselage arrives at Weybridge on a special trailer. (BAe Systems Heritage)

and systems bays were immediately behind the cockpit, while the twin-wheel nose undercarriage retracted rearwards into the bay below the No. 1 fuel tank. The cockpits and radar and main equipment bays were pressurised and environmentally conditioned.

The crew were provided with ejection seats and the pilot had a head-up display (a novelty back then), which gave him all essential flight and attack information in the form of symbols focused at infinity on the windscreen. The windscreen was capable of withstanding a one-pound bird impact at transonic speed, unlike its later counterpart the American General Dynamics F-111, which had no bird strike requirement written into the specification.

The forward centre fuselage was relatively short and was flanked by the intakes. Its upper region was occupied by part of the No. 2 fuel tank and its lower region by bays for the Doppler radar, inertial navigator, retractable ram-air turbine to generate AC electrical power in an emergency and other systems.

The much longer aft centre section contained the remainder of the No. 2 fuel tank, which lay between and over the engine intake ducts. The rear fuselage housed the twin Olympus engines, the jet pipe tunnels for which formed an integral part of the fuselage. An 80-gallon water tank, which enabled water to be sprayed into the engines' combustion chambers to restore thrust for operations in hot conditions, was fitted between the jet pipes. An airborne auxiliary powerplant was fitted in the fuselage forward of the weapons bay and this provided low-pressure air for engine starting, AC electrical power for fuel pumps, throttles and the inertial platform, ground running of the main engine gearboxes and air-conditioning. Four self-contained hydraulic systems were to feed duplicate power controls, and electrical power came from two engine-driven generators, while air tapped from the engine compressors provided pressurisation and air-conditioning.

The rear fuselage was mainly occupied by the Nos 3 and 4 fuel tanks, which almost completely surrounded the power units. The spigots, about which the fin and tailplanes were rotated, were attached to a frame that also served as the rear bulkhead for the No 4 tank. The rear fuselage was later redesigned to include the whole of the engine installation. There were three airbrakes on the rear fuselage, with one on each upper shoulder and the third centrally on the lower surface. The fuselage was completed by a detachable fairing that provided stowage for the brake parachute and formed a shroud for the jet pipes' ejector nozzles.

The main undercarriage and engine accessories bay were located in the lower fuselage, flanking the centrally positioned weapons bay that ran the whole length of the section. Each leg of the forward-retracting main undercarriage carried a pair of tandem-staggered large-diameter low-pressure wheels mounted on a bogie beam that was rotated to align with the leg before retraction. The large main doors covering the stowed undercarriage were designed to close again after the mainwheels

A complete TSR2 wing section leaves Samlesbury for Weybridge. The TSR2 wing area was 702.9sq ft and the entire wing was an integral fuel tank. (BAe Systems Heritage)

had been lowered. This double cycle arrangement, which was later adopted for the nosewheel doors, was intended to reduce the loss of lift and increase in drag if the bays had been open during takeoff and landing. The twin bogey undercarriage installation absorbed much design effort as it had to cope with an emergency landing at maximum takeoff weight and, with low-pressure tyres, allow operation from unpaved surfaces. For short takeoffs the nose could be raised by extending the nose leg shock absorber strut.

The TSR2 could use AVTAG, AVTUR or AVCAT fuels, and dual connections enabled a high refuelling rate of 450 gallons per minute. An in-flight refuelling nose probe was to be fitted.

Fuselage and wing fuel tankage totalled 5,588 gallons. Underwing fuel tanks plus an extra tank in the weapons bay for long-range ferry flights led to the aircraft being considered for the tanker role.

The thin delta wing was of multi-web construction, with integrally stiffened skin panels, the primary structure containing two integral fuel tanks. A few ribs were positioned to diffuse loads from the trailing-edge flap hinges and external stores and wing/fuselage attachment points. The outer two-thirds of the leading-edge were extended forwards of the main wing, the inner half of the extension being made into a plain flap, creating a saw-tooth at its inboard end and a notch as its junction with the fixed outer portion. The flap was to assist in providing stability at low speeds when the main flaps were deployed more than 15 degrees. The innermost part of the leading-edge was fixed and incorporated a high-frequency radio notch aerial at each apex. The whole free length of the trailing-edge was occupied by plain tapered blown flaps.

The low-aspect ratio delta tailplane was low set to provide combined aileron and elevator functions. Plain trailing-edge flaps fitted to the tailplanes were geared to their movement and to blown-flap deployment to increase tailplane effectiveness and to reduce their range of travel at low speeds. The outline specification had sought short field operation to allow the TSR2 to take off from half a runway, the other half having been bombed. It was at this stage that the requirements were extended to include operation from 'disused' airfields and rough strips generally. A standard rough strip was eventually devised, based on measurements of old European and Middle East airfields.

For this all-important STOL performance the full-span flaps (from root to anhedral tips) used blown air from the engine compressors. The flaps were automatically locked in the neutral position at high speed. The aircraft had an all-moving fin rather than a conventional fin and rudder. The all-moving tailplane sections moved in unison to provide longitudinal control, or differentially for lateral control.

The most apparent changes made to the layout of the TSR2 during 1960 were the adoption of variable half-cone intakes and distinctive wingtips with pronounced anhedral, and the abandonment of leading-edge high-lift and airflow control devices. Other changes included a nose undercarriage leg designed to extend during the takeoff run to relieve the lift forces on the tailplane sections and to reduce the speed at which the nosewheel began to lift, and thus shorten the takeoff run. Four airbrakes, instead of three, were fitted to each of the upper and lower shoulders of the fuselage.

Unfortunately, the more that the TSR2 evolved the more 'Empire building' on a grand scale became the order of the day. Everyone in the Ministry of Defence (MoD), the Ministry of Aviation (MoA) and industry wanted a piece of the action, and it became commonplace for as many as 60 people to sit down to discuss a technical problem for hours on end without result. There was a Management Board, Systems Integration Panel, Sub-System Committees and Working Parties to whom decisions were referred, and they reported back and then reconsidered.

A classic example was the TSR2's fuel system. BAC evolved a down-to-earth capacity gauging system based on experience with the Scimitar naval attack aircraft. This involved two systems in parallel with the pilot switching between tanks and the navigator doing the manual trimming. EE favoured a flow-meter system for balancing fuel usage between the tanks. The MoA led the relevant Committee, which decided eventually on the adoption of gauging, flow metering, and an automatic comparator. The fuel system was two or three times as complex as it need have been, with a corresponding increase in price.

During the ongoing design process much time was expended in trying to cater for all situations in which the aircraft might find itself. For instance, to meet the requirement that the TSR2 would have to takeoff from any one of a hundred small flying fields in Europe, BAC produced detailed plans for a 'General Servicing Vehicle' (GSV). A four-wheeled multi-fuel-engined air-conditioned vehicle, GSV was to tow the TSR2 to a refuelling area, provide power for refuelling and water replenishment, electrics and air-conditioning and replenish and function-test the hydraulics. It was estimated that the vehicle could maintain the TSR2 at instant readiness for periods of up to 30 days. The GSV was to be air-transportable in an Argosy freighter.

PERFORMANCE COMPARISONS								
	Engine thrust (lb)	Weight, empty (lb)	Max weight loaded (lb)	Max speed sea level	Max speed altitude	Initial rate of climb	Radius of action (nm)	Ferry range (nm)
Canberra B(I)8	2x 7,500	23,170	56,250	450kts	Mach 0.88	3,600ft/min	700	2,940
TSR2	2x 19,600: 30,600 with reheat	54,750	103,500	800kts	Mach 2.25	15,000ft/min	1,000	3,000+
Tornado GR1	2x 9,980: 17,270 with reheat	30,620	62,832	800kts	Mach 2.20	15,000ft/min	810	2,400

CHAPTER FOUR

ENGINES AND AVIONICS

The choice of the TSR2's engines, and their positioning, was just one of the major questions facing EE/Vickers in the beginning. Sir George Edwards, Managing Director of Vickers, who was deeply involved at the time with the VC10 airliner, wanted to hang them externally under the TSR2's wings or fuselage, rather than integrate them into the basic structure where serious problems such as the use of cooling air and multiple heat shields were bound to arise. Indeed, there were slighting Vickers' references to engines in fuselages giving rise to a 'hot can of worms'. In the end it was decided to put the engines in the rear fuselage in a similar installation to that on the original P17.

Both Rolls-Royce and Bristol Siddeley engines were assessed for the TSR2, and it was unanimously decided that the former was best placed to solve the many problems associated with high supersonic speeds and best specific fuel consumption at low altitude. However, the MoA refused to accept the recommendation and BAC was told that the TSR2 was to be powered by two Bristol Siddeley Olympus B.O1.22R twin spool jet engines. Each 22R engine, equipped with water injection, was to give 19,600lb dry thrust, sea level static, and 30,610lb with maximum reheat. They were derived directly from the Olympus B.O1.15R by the addition of a zero-stage to the low-pressure compressor, and had a fully variable reheat system, the simple convergent exhaust nozzle progressively opening with increasing reheat temperature.

The punchy thrust available from these engines was deemed essential to ensure good takeoff characteristics and, in certain circumstances, to give the TSR2 a better rate of climb than the Lightning interceptor.

The TSR2 final assembly jigs in October 1961. Optical equipment was used to precisely align each jig in sequence. (BAe Systems Heritage)

The port engine half-cone intake being built. The half cone engine intakes were adopted in mid-1959. (BAe Systems Heritage)

Each TSR2 engine was controlled automatically by an electrical system using a conventional throttle box. The ideal engine intakes were of the variable double-wedge type, located ahead of the wing.

John Wragg, who joined the Engine Division of the Bristol Aeroplane Company in 1952 as a development engineer, recalled how the company dealt with some of the challenges associated with producing the engine. 'I was appointed at the time of the launch of the TSR2 engine as the project engineer in charge of the development programme, and I was responsible for the introduction of design changes to resolve the various problems that were encountered.

'When it was announced in January 1959 that the Olympus turbojet was to power the TSR2, the engine had already been in service with the RAF powering the Vulcan B Mk 1. The first major change in design was the Mk 201 of 17,000lb developed to power the Vulcan B Mk 2. A second re-design occurred with the B.O1.21 Mk 301 of 20,000lb thrust, also used in the Vulcan B Mk 2. The B.O1.22R Mk 320 engine for the TSR2 stemmed directly from the B.O1.21. The Olympus 593 engine that powered Concorde evolved from the 22R and was the seventh member of the family.

'The Olympus was Britain's first twin-spool turbojet, and this configuration underpinned the engine's outstanding handling characteristics and exceptionally low specific fuel consumption. Being one of the most rugged and reliable engines in service with the RAF, the valuable operating experience accumulated in service with Bomber Command was an important consideration when the Olympus was chosen for the TSR2.

'From an engine manufacturer's point of view, the TSR2 was every man's incentive in terms of the development of a gas turbine engine. It was demanding in every sense. It was seeking things that had never been

CHAPTER FOUR **ENGINES AND AVIONICS**

done by an engine before and all of these were naturally pursued with vigour, but perhaps not early enough, and the preparatory work was not done successfully enough to be able to say that the engine would be fully developed at the time when it was required for production aircraft.

'The reheat system for the engine emerged from work that had been done with a Solar reheat system developed partly in the United States, and partly at Patchway [in Gloucestershire, home to the Bristol Aeroplane Company's Engine Division] on various Olympus engines. An infinitely variable reheat system, combined with a variable nozzle that was later pneumatically operated, was the basic standard developed for the TSR2, and was in its successful demonstration the basic reason why this particular engine was selected and why that particular reheat system was chosen. Infinitely variable reheat systems are quite difficult to design and develop and of course they need a variable nozzle of a degree of complexity. So that was one step forward, the thrust required to make the aircraft go supersonically being something which had been demonstrated.

'A great deal of attention was also focussed on what was required of an engine that needed to undergo sustained operation at high Mach numbers, creating an intake total temperature of 146°C (Concorde demanded 127°C). Much of the design to do this had to be attacked on an analytical basis as there was no evidence that could be looked at

The TSR2's starboard Olympus engine. The engines were a very tight fit within the airframe. (BAe Systems Heritage)

and, of course, the Olympus engine designed for the Vulcan did not have to deal with this sort of condition. The fact that air, oil and fuel temperatures were much higher in the TSR2 was a new challenge and demanded a changed construction and new materials within the engine. The new arrangements for the shafting to move bearings out of the very hot areas of the engine were more or less adopted off the drawing board. Although analysis work was done to identify that with the design changes throughout the engine it should be able to operate at the much higher temperatures that were required for supersonic flight, it could not be demonstrated until the first Olympus 22R bench engine had run.

'Testing the Olympus engine for the TSR2 necessitated a new schedule with many new mandatory additional tests. The Committee sizes and structure for deciding on the overall requirements of the production of the TSR2 would today be unbelievable. It was as if we were to say to everybody today at the launch of an IT system "what would you like to have out of this system" and there was nobody who ever said "no, you cannot have that". On some of the most difficult elements of the engine, the weasel wording was by today's standards absolutely appalling. The phrase used was that the engine should be "compatible with the aircraft intake". But what standard is going to be met and by whom? There was no mention of what testing had to be done to demonstrate that the combination was adequate.

'The worst problem that dogged the engine for the TSR2 in its development programme was the behaviour of the low pressure [LP] shaft. A much longer LP shaft was fitted to the 22R than had been in use on earlier Olympus engines, and that was, in part at least, as a result of trying to keep the engine bearing compartments as cool as possible in the much hotter environment that was going to be encountered in supersonic flying. But unfortunately that change also resulted in a shaft design which was capable of being excited in vibration by a number of stimuli, one of the most significant being the resonance of a shaft mode with over-fuelling of the reheat system – this had been discovered immediately prior to the first flight of the TSR2. It was necessary to revise the fuelling of the reheat system at that stage, and subsequently to introduce a completely new schedule to ensure that the over-fuelling and the consequent excitation were avoided. That was done and was successful, but there were many flights that were carried out before those changes had been fully introduced.

'How would I sum up the experience of developing the engine for the TSR2? Well, it achieved delivery of an engine, deficient in a number of areas I agree, but adequate for early TSR2 flying. Subsequent development of Olympus for Concorde shows that there was still very much to be discovered at that stage and the cancellation of the TSR2 prevented the Olympus 22R engine being developed to match fully the requirement of the aircraft. The first engine to benefit from a full, combined customer/industry funded, demonstrator programme was the EJ200 for Eurofighter, and all the evidence points to an extremely successful validation programme as a result.'

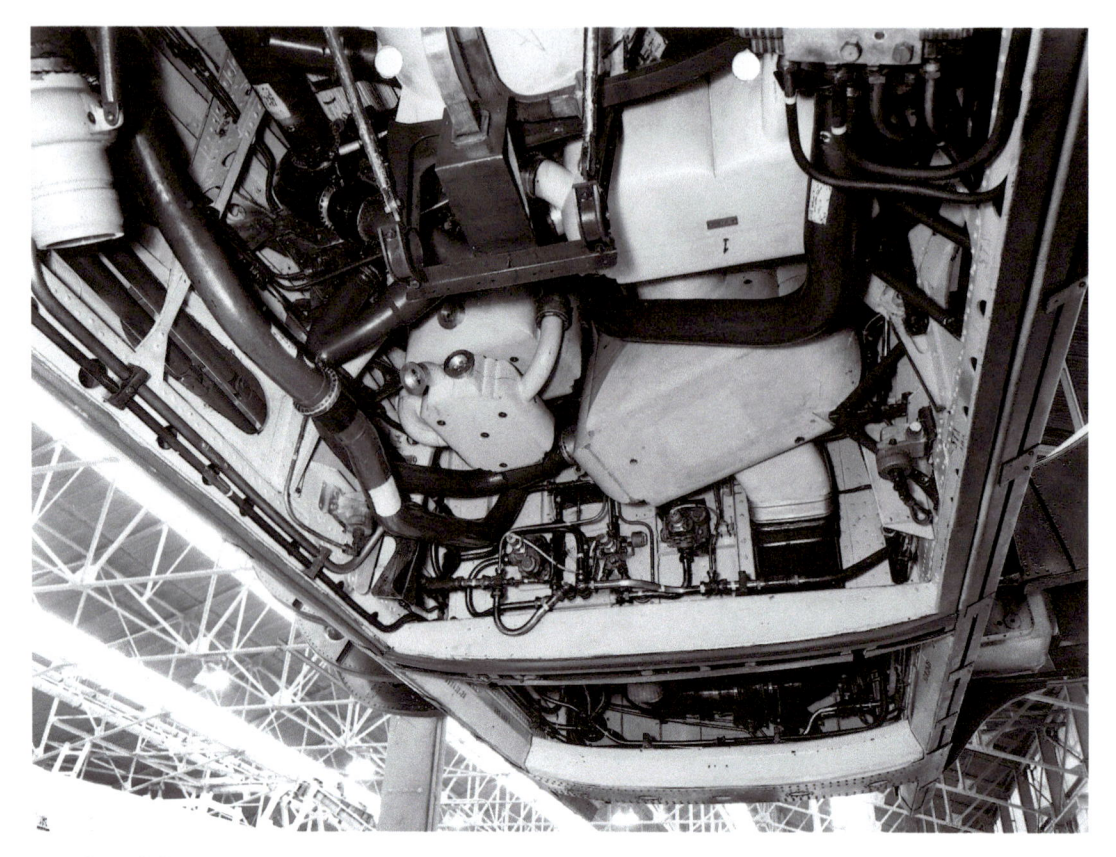

CHAPTER FOUR **ENGINES AND AVIONICS**

ABOVE The TSR2 equipment bay being built. (BAe Systems Heritage)

AVIONICS

The ability of the TSR2 navigation and bombing system to operate for long periods without giving the game away was an important attribute for deep penetration into hostile airspace.

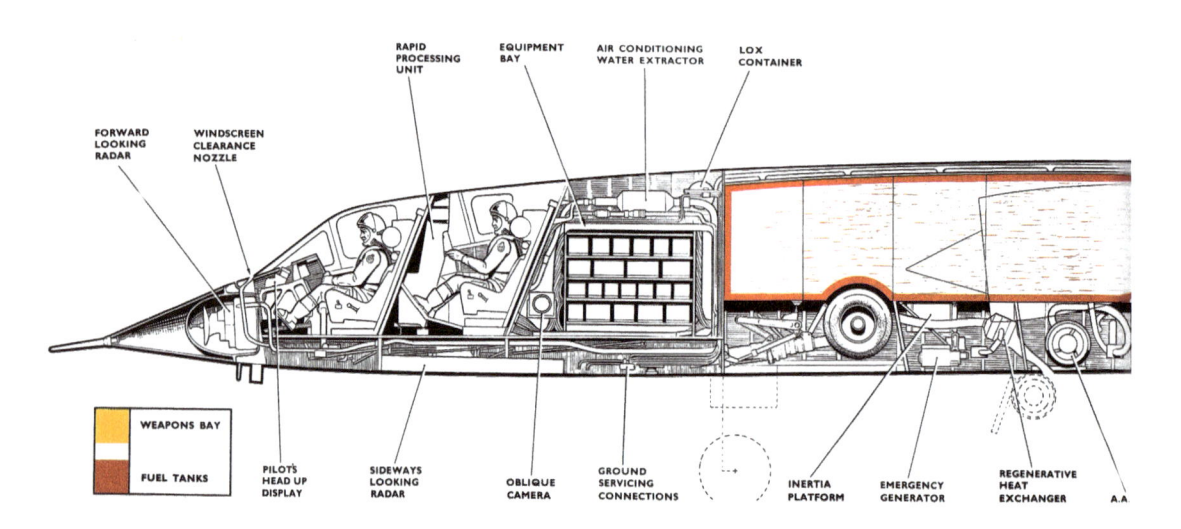

The TSR2's navigation attack system used forward and sideways-looking radars, Doppler radar and an inertial navigator. The primary function of the forward-looking radar (FLR) then under development by Ferranti was terrain following, but it was also used for homing and for weapon guidance and ranging. The Doppler measured ground speed and drift angle to a very high degree of accuracy throughout the aircraft's flight envelope, supplying inputs to a display in the navigator's cockpit and to the navigation and bombing computer.

Directional information was provided by a Doppler-inertial mixed dead-reckoning system. Position readings from the two units were fed into a central computer, which checked them against a predetermined course. Any deviations from the flightpath were calculated by the computer and the corrections obtained from it fed to the autopilot, which returned the aircraft to its original course. Navigational errors accrued over long distances were periodically checked at predetermined points on the flightpath by the SLR, which provided the navigator with a picture of the local terrain to pinpoint his position exactly and to update the aircraft's coordinates directly to the navigation computer. The SLR could provide information in blind conditions at high and low altitudes. In addition, a photographic record, which was used for reconnaissance purposes and the correction of maps, could be obtained from the radar display. The two aerials of this radar were located on either side of the aircraft below the cockpit floor.

The four main TSR2 operating modes were envisaged as follows:

1. Long distance operation, navigating without external aids.
2. High speed flight at high and low levels (down to at least 200ft), with minimum exposure to enemy ground fire.
3. Reconnaissance to record details of ground and man-made features over a wide swathe centred on aircraft track.
4. Accurate delivery of ordnance to predetermined targets and targets of opportunity.

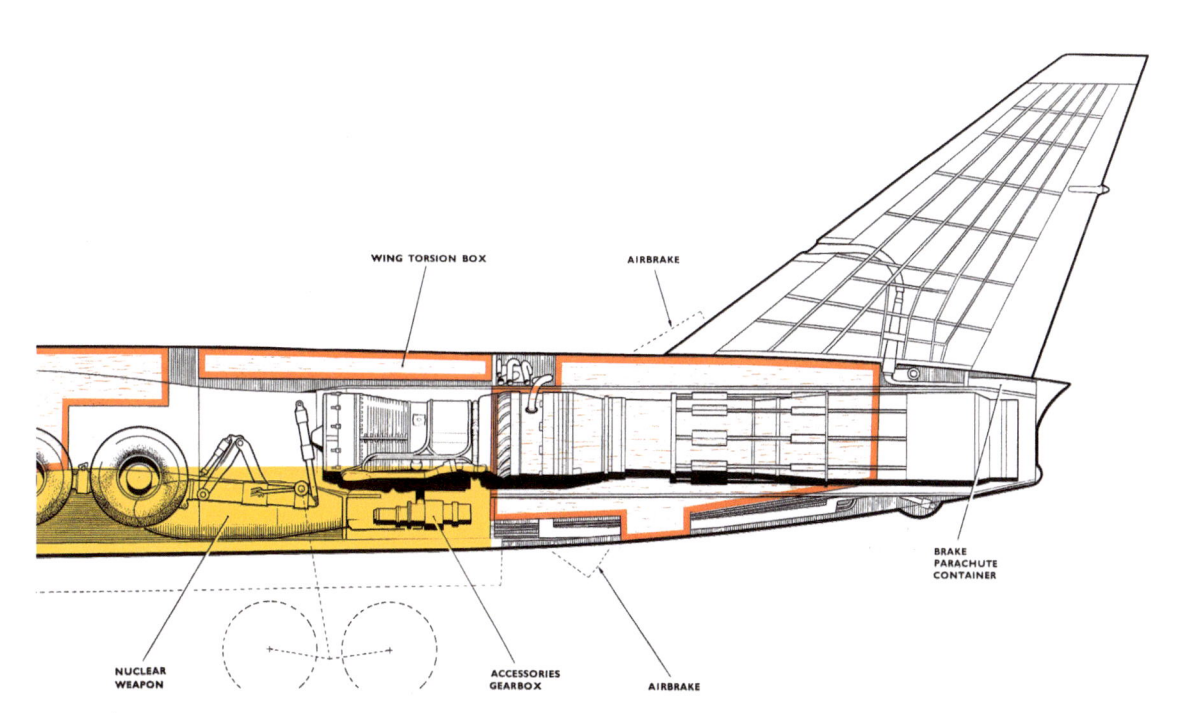

BELOW AND LEFT Cutaway of the TSR2's fuselage. (Brooklands Museum)

The principal equipment to meet these requirements were assessed as:

EQUIPMENT	SUPPLIER	LATER
1. Forward-Looking Radar (terrain-following)	Ferranti	GEC-Marconi Avionics
2. Inertial Navigator	Ferranti	GEC-Marconi Avionics
3. Doppler Radar	Decca	Racal-Thorn
4. Central Computer (Verdan)	Elliott Bros	GEC-Marconi Avionics
5. Moving Map	Ferranti	GEC-Marconi Avionics
6. Radio Altimeter	STC	STC
7. Head-Up Display	Rank Cintel	GEC-Marconi Avionics
8. Flight Control System	Elliott Bros	GEC-Marconi Avionics
9. VHF/UHF Radios	Plessey and Marconi	GEC-Marconi Avionics
10. Sideways-Looking Array Radar (SLAR) – X-Band Navigation Radar	EMI	Racal-Thorn
11. Sideways-Looking – Q-Band Reconnaissance Radar	EMI	Racal-Thorn
12. Optical Line Scan	EMI	Racal-Thorn
13. Stores Management	Vickers Armstrongs (Weybridge)	British Aerospace
14. Head-Down Display	Plessey and Marconi	GEC-Marconi Avionics

LOW-LEVEL FLIGHT

The key to safe low flying over varying terrain in all weathers with minimum exposure to enemy surface engagement lay in determining the ground contours along the aircraft's flightpath, and providing steering control signals in the vertical plane to enable an optimum profile to be flown. With the TSR2 this was to be done via a terrain-following radar (called the FLR to differentiate it from the other radars carried). It operated as part of a system that included a radio (as against the modern radar) altimeter, measuring the distance to the ground or water directly below the aircraft, angular stabilisation signals from the inertial navigator, an airstream direction detector (ADD) to determine the aircraft's velocity vector in the vertical plane and azimuth drift angle, once again from the navigation system, to ensure that the radar scanned the ground along the aircraft's future track. The earliest trials of this system, in a de Havilland Comet, were encouraging, with performances of around 6–8nm per hour circular error probability in position and about 0.25 degrees per hour in azimuth.

The intention was to couple the steering command signals computed by the radar to the Automatic Flight Control System (AFCS), with a Head-Up Display (HUD) providing the pilot with tracking performance symbology. However, manual control using only the HUD was also possible – this was used during flight trials, which initially utilised a Dakota, then a Canberra and finally a Buccaneer in extensive flying over the rugged Scottish Highlands.

The terrain-following radar system provided safety in turns by arranging for the radar's vertical scan to lean into the turn to cover the

A hive of activity as the prototype undergoes final assembly. The relative size of the assemblers shows that the TSR2 was a big beast. The fuselage was only 8ft shorter than that of the Avro Vulcan. (BAe Systems Heritage)

correct area of ground along the future track. Although it was never possible to test the whole system with the AFCS, since the trials aircraft did not have a suitable one installed, many miles of manual flight over very rough terrain at heights down to 98ft without hazarding the aircraft were a testimony to its success. This was probably the first time that a radar had been so closely linked to the flight control of the aircraft.

The FLR contributed to other operational roles including ground mapping for navigation and target identification and beacon homing to rendezvous with tanker aircraft. With a pre-planned target or one selected in flight by the navigator, transition to the attack mode from terrain-following was accomplished by automatically injecting a climb command into the vertical steering signals, which induced a gentle bunt to achieve the right position for visual acquisition of the target by the pilot. By flying to place the HUD aiming mark over the desired target on the ground, the radar, with its boresight slaved to the aiming mark, could then measure the range along this boresight to the intersection with the ground. From this range and range data, the weapon release point could be computed.

The display for the FLR was located in the rear cockpit. To offset the expected high turbulence that would have made it difficult and tiring for the navigator to focus on the display, the CRT was viewed

CHAPTER FOUR ENGINES AND AVIONICS

1970s TSR2

This three-view shows the TSR2 in service with No. 40 Sqn in low-level camouflage synonymous with aircraft flown by Strike Command in the 1970s. This squadron, disbanded in 1957 after operating the Canberra, was expected to be the first operational unit to receive TSR2s.

indirectly via a lens and a mirror that combined to form a long folded optical path. This made the image on the display tube appear to be at infinity, and therefore quite steady to the observer, despite any ambient vibration of the display or himself.

RADAR NAVIGATION AND RECONNAISSANCE

The FLR and the Doppler radar were not the only radars installed in the TSR2. An X-band EMI SLR, utilising two long arrays (one on either side of the aircraft), was to present a detailed view of the ground and to allow 'fixes' to be obtained that could correct the prime inertial navigator. The transmissions were switched alternatively from one array to the other to give a complete ground picture centred on aircraft track. This was to be presented to the observer by an unusual display system called the Rapid Processor.

Since SLRs utilise the forward motion of the aircraft to scan the ground, the picture is built up by adjoining strips or lines on the CRT

The TSR2 avionics test rig (BAe Systems Heritage)

display, each representing the returns from an individual or group of transmit pulses. These lines must be integrated continuously to show the complete map, and on the TSR2 this was to be done by exposing the lines on the CRT to a photo-sensitive strip of film that was being moved past it and then instantaneously developed using a series of chemicals contained in bottles attached to the display. The complexity of the mechanisms that must have been required to operate this system successfully in the ambient conditions of the cockpit was remarkable.

A further navigation aid was a projected Moving Map Display, also from Ferranti. This held a large area of Europe on 35mm film and was driven by the Navigation Computer to present the aircraft's current location in the centre of a ground glass screen. Zoom and look forward facilities were available by selection. A repeat display was provided in the front cockpit. In its initial form the image projected by the Moving Map Display was created independently of the aircraft's radars, but later developments allowed these to be combined so the map overlaid the radar display and made it easier to recognise ground features and update the navigation system by bringing the two displays into coincidence. Again, this might well have been a future TSR2 upgrade.

To gather data in the reconnaissance role, a special underbelly pack was to be provided. This contained, besides a bevy of cameras, a very high resolution SLR operating in Q-band (37GHz) and an optical line scan unit. These were both developed by EMI. Outputs from these units were to be held on a separate photographic recorder, and

CHAPTER FOUR **ENGINES AND AVIONICS**

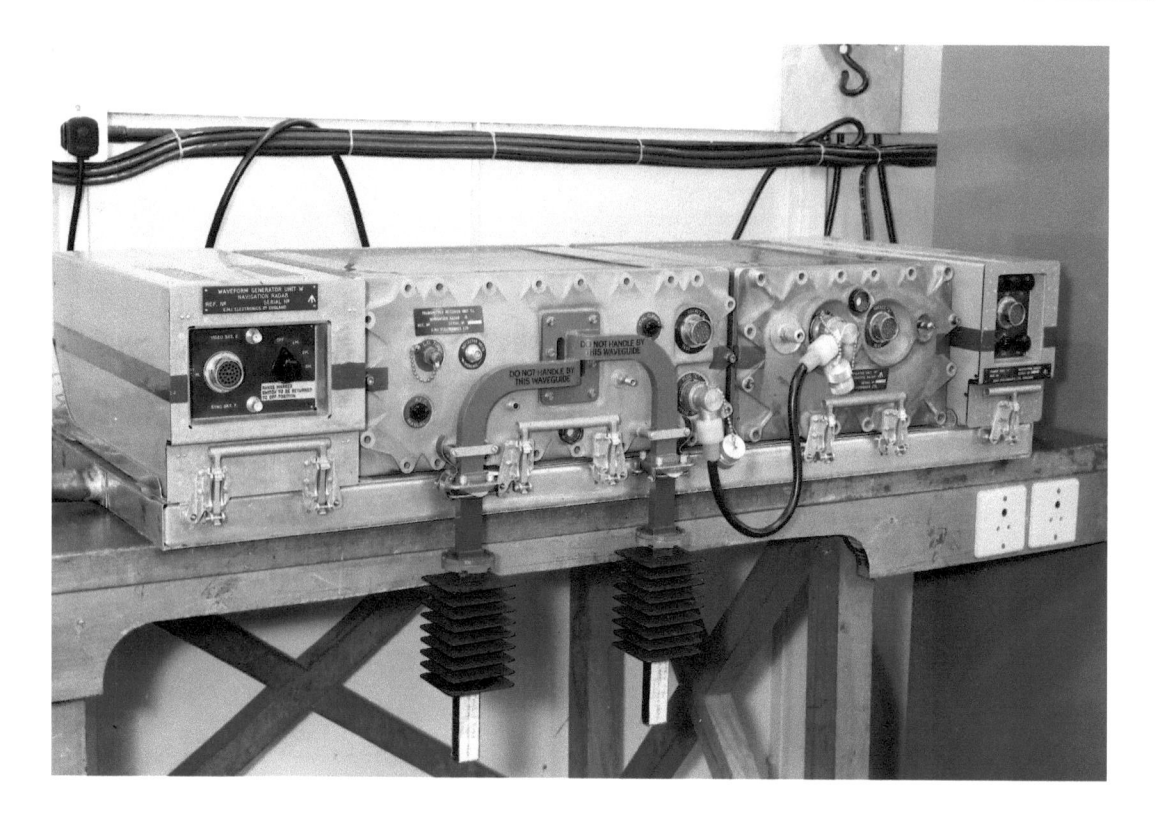

The TSR2's SLR on the test bench. This comprised the waveform generator, transmitter/receiver, modulator and power unit. The SLR aerial was a seriously long piece of kit. (Brooklands Museum)

the creation of a ground processing and replay facility that was to have data relayed by video link from the aircraft in flight was also looked into. In an age before satellite communication it is difficult to see how this broadband link was to be maintained over long distances and with possible terrain masking, but the idea was right.

Sir Donald McCallum joined Ferranti Ltd in 1947. The three major programmes that Ferranti was involved with for the TSR2 were the development at Bracknell, in Berkshire, of the Inertial Platform designed by the Royal Aircraft Establishment (RAE) and the design and development of the FLR and the Moving Map Displays in Edinburgh. Until early 1968 Sir Donald was Manager of the Electronic Systems Department in Scotland responsible for the last two of these projects.

'Before concentrating on the FLR and especially on the terrain-following mode, one aspect of the Moving Map Display deserves at least a mention. Later the computer was modified slightly to become the attack computer for the Nimrod and continued unchanged into the Nimrod MR2. Even in 1962 it was extremely old-fashioned in its technology – ball and disc resolvers, and not a digit in sight. Maybe there is a moral somewhere.

'In 1957 I went to a meeting at Warton to discuss missile integration for the Lightning with Hawker Siddeley Dynamics. Their aircraft was delayed several hours and in the waiting period Tony Simmons told us of some of English Electric's forward thinking. The best defence for future strike aircraft seemed to be in very low level flight where advantage could be taken of hills and valleys to provide concealment

First prototype TSR2 XR219 in final assembly at Weybridge on 31 August 1963. The 30-degree down-turned wingtips were added in early 1960 to counter Dutch rolling. (BAe Systems Heritage)

and the task of the guided weapon made technically more difficult because of ground clutter. However, it was clear that such a path could not be flown visually, certainly not in the night and bad weather conditions which gave extra security.

'Simmons then told us of a proposal for a radar system scanning in the vertical plane and using Q-band (0.8cm wavelength) to get adequate resolution. Immediately we both exclaimed that the work Ferranti had done on monopulse (static split) air-to-surface ranging at X-band (3cm wavelength) would give more accurate information and reduce the weather problems inherent with Q-band. Within a few minutes we had outlined the so-called angle-tracking system in which the radar tracks the ground ahead of the aircraft and determines by measurement of the difference between this angle and the aircraft's flight vector the appropriate flightpath.

'Very little happened for some time after this until a meeting was held in, I think, 1960 to discuss the problems of low flying for the TSR2. We put forward our proposal and the representative of the Royal Radar Establishment described their preferred solution, which involved a pre-planned tape recording of heights along the planned route, with voice instructions to the pilot on change of height. It found little favour, especially with the pilots present.

'The Director of Electronic Research and Development (Air) in the MoA then provided funding for work to be done on the Ferranti proposal. A detailed system study was undertaken and existing information on radar returns from ground surface and obstacles was

CHAPTER FOUR ENGINES AND AVIONICS

Tail section final assembly. The TSR2's tail-braking parachute was stored in the bay above and between the jet pipes. (BAe Systems Heritage)

studied. Accurate information proved to be very scanty, in fact virtually non-existent, and a programme to establish quantified information had to be established. This involved detailed measurements from a ground site at Linlithgow, near Edinburgh, which included a wide range of targets, including a palace, and flight observations from a Dakota to determine the type of echo from, *inter alia*, television towers and power lines. These showed, in the case of trees, for instance, that while some echoes appeared to be from below ground level, there were always echoes during a series of a few pulses from the top or higher than the actual tree. This gave confidence to continue with the system design.

'The FLR used many components from the AI23 and AI23B radars fitted to the Lightning, including the mechanically roll-stabilised scanner. This was to prove of great value in turning flight. The transmitter power was reduced from 250kW in the AI series to 50kW, which was still considerably higher than competing equipment and a considerable safety advantage at very low levels of flight. The electronic circuits were, of necessity, a new design using solid state devices.

'In our previous experiences on the Lightning and the Buccaneer the aim had clearly been to develop and produce a total system to meet the operational requirement. In both of these aircraft there was no question of who was responsible for the overall design, Freddie Page on the Lightning and Barry Laight on the Buccaneer. In the case of the TSR2 this was never clear to me. The enthusiasm which the earlier programmes generated was absent. The "system concept" meetings isolated us from the aircraft and I never saw a TSR2 airframe during the entire programme. These meetings had an unreal air, perhaps in line with the dictionary definition of concept as "an abstract idea". Initially, the meetings were chaired by George Henson and were long on discussion but short on decision. For instance, there was a problem with transients on the 115-volt three-phase 400 Hertz supply which took many man-hours of talk on the possible causes of such an annoying problem when all that was needed was for the designers of the power supply to be told to fix it, which they eventually did.

'In the last year or so of the programme Sir George Edwards put Harry Zeffart in charge of equipment. "Twenty-eight volt Harry", as the system concept men called him, quickly brought realism to the situation. Having been through the fire of getting the Valiant and other aircraft into service, he knew the hard grind that is essential, and that repeating the words "system concept" as a kind of mantra was no substitute for action. His appointment two years earlier would have transformed the equipment situation.

'At 0200hrs on 6 April 1965 Ferranti engineers completed testing the first FLR installation in a TSR2, and in the afternoon of the same day James Callaghan, Chancellor of the Exchequer, announced the cancellation of the TSR2.

'The whole FLR programme reflects great credit on the team led by Dick Starling and Greg Stewart and on our chief test pilot, Len Houston, who flew the flight trials manually using the steering information displayed in the HUD. When the Buccaneer flight trials were completed and analysed and the results shown to Frank Pelton, the pioneer of terrain-following at Cornell, he told us our achieved performance – where we had demonstrated flights at 100ft over the Scottish Highlands compared to the design target of 200ft – was several years ahead of the rival work in the States.'

AUTOMATIC FLIGHT CONTROL SYSTEM

The AFCS provided three-axis autostabilisation involving control of both fin and tailerons, plus autopilot/flight director control of the aircraft in azimuth and elevation. The AFCS was one of the few systems to be tested, in part, on board the TSR2 aircraft itself. By the time the project was cancelled in 1965, extensive ground testing of the structural feedback from the rate gyros and accelerometers had been carried out to verify frequencies and node positions. That said, little flight experience had been obtained with the many novel features of the AFCS.

Much of the TSR2 experience was used in the design of the AFCS for Concorde, by which time integrated circuit DC amplifiers were available to enable a major reduction in electronic component count. The triplex actuator concept was further developed by Elliott Bros in a quadruplex configuration that was the basis of the Tornado electro-hydraulic actuator. Concorde and Tornado were perhaps the last systems to benefit from a continuity of experience from the Lightning, Buccaneer, VC10, BAC 111 and TSR2, all these projects being undertaken in not much more than a decade by the Elliott Bros team, with many individual engineers sharing their experience freely between the projects.

COMPUTING POWER

By the time the TSR2 project was terminated, most of the avionics equipment was in an advanced stage of development, and probably all of it had been subjected to flight trials, although a complete avionics system was never flown in the TSR2 itself. Much equipment, however, including the FLR, had been installed in the No. 3 TSR2 prior to flight test when the end came.

TSR2 was particularly ambitious given 1960s avionics technology. The potential advantages of semi-conductors were readily appreciated but the integrated circuit was a relative novelty incorporating perhaps a few transistors against the million or more in 21st century processor chips. Avionics back then were predominantly analogue, and the most obvious difference between the TSR2 and today was the absence of the digital bus as the primary means of transferring data and control signals between individual avionics equipment. Instead, individual wires, each with a single purpose, led to large and heavy cable forms running the length and breadth of the aircraft with, perhaps even worse, an associated large number of multiway plugs and sockets.

Perhaps the most novel feature of the TSR2 avionic system, and the one which proved hardest to implement, was the decision to use a pair of digital computers as the aircraft's central means of carrying out the navigation and bombing calculations. Accurately determining present position and track placed heavy emphasis on the reliability and accuracy of the navigation system. This was to be met by the then latest technique – a Doppler-Inertial mixed system which, while keeping the platform inertial velocities correct, helped keep the platform to the local vertical. The first stage of the computation (converting acceleration to velocity) was to be accomplished

The TSR2's starboard avionics bay being equipped in May 1963. (BAe Systems Heritage)

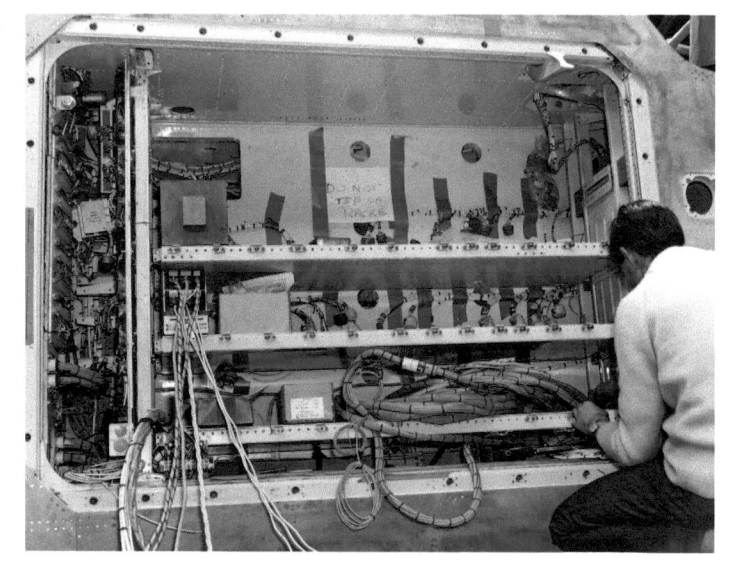

Pre-micro miniaturisation – the TSR2's Verdan computer assembly. The casing for the memory disc stands out, as does the large hose providing cooling air. (Brooklands Museum)

by electromechanical integrators but the second stage (velocity to position) was to be the responsibility of the central digital computer.

Digital computers had not been used previously on British military aircraft, but the need for precision calculations in inertial navigation systems and the increasing difficulties of combining large numbers of different analogue electronic subsystems in an overall integrated multi-function system had become very apparent during the development of the Navigation Bombing System (NBS) Mk I for the Victor and Vulcan Mk 2s and their Blue Steel stand-off ballistic missile inertial navigation system.

Such difficulties led to the TSR2 specification for a system based on a central digital computer. This would give precision and improved capabilities in the integration and multifunction aspects of the system. GEC Stanmore, in association with RAE Farnborough, had developed a prototype digital computer/differential analyser (called Dexan), but it was far from mature and lacked the necessary software development infrastructure, even for the 'small' (by modern standards) amount of programming required – some 4K to 8K words. It is worth noting that 90–95 per cent of the TSR2 development effort was devoted to hardware-related matters, most of which involved entirely new and unfamiliar technologies, and five to ten per cent to software matters. For the F-35 Lightning II, it is probably five per cent hardware to 95 per cent software.

Although the RAE preferred its Dexan machine, Elliott Bros had considerably greater overall knowledge of digital computing. It was the prime contractor for the Blue Steel inertial navigator and the NBS aircraft interfaces. Combining this background with an aggressive campaign to licence and market the North American Aviation Verdan (Versatile Digital Analyser) computer, Elliott Bros won the contract for the central digital computing system – the first in a European aircraft.

The two Verdan computers installed in the TSR2 were virtually the only non-British avionics in the aeroplane. Verdan had originally been designed to carry out inertial navigation calculations and platform control for the earliest US Navy nuclear submarine, USS *Nautilus* (SSN-571), and for the Navajo cruise missile. It was later adopted for the Hound Dog cruise missile and for the North American Aviation A3J Vigilante's integrated navigation system, where it performed a central computer task not unlike but considerably less complex than for the TSR2.

The Verdan computers operated in serial mode and had very limited capability compared with even the most modest PC today. Even so, although powerful for their time, it quickly became evident that the computers' capacity to satisfy all the demands placed on them were distinctly limited. This was aggravated by the fact that the remainder of the avionics system was analogue, and, therefore, inputs and outputs had to be transformed through A-to-D and D-to-A converters, creating a bottle-neck and great competition for the available resource.

Peter Hearne retired as Chairman of GEC Avionics in 1994. Earlier, as an engineer, manager and later director of Elliott Flight Automation, he played a major part in the development of computer systems for the TSR2 and other aircraft. 'The reliance of the TSR2's systems on computer power was just one of the novel and demanding aspects of the project. Probably the most difficult design task was the Navigator's Control and Display Unit (NCDU). We had been allocated a space about 18 inches high at the right hand side of the cockpit spanning the junction between the upper and lower sections of the cockpit structure. The mechanical design of the unit was a nightmare. The box had some 12 faces, of which only two met at right angles and at least one was curved – this was in the days before computer-aided design. The fixings provided for the unit were at its base, with a rear-mounted connector to allow it to be jacked in and out for maintenance purposes. These fixings proved to be inadequate for a unit of its size and weight, so a top fixing was added.

'Design of the NCDU began around 1960 using germanium transistors [also used by Verdan]. Silicon transistors did not become generally available until a year or so later, and I recall an agonising decision as to whether we should change to the new silicon components, which meant changing the polarity of all the circuits we had designed. I am pleased to say we took the right decision (for once).

'The central computing system was required to carry out all of the special weapon (aka nuclear) bomb delivery calculations and all of the outer loop navigation calculations, together with the integration of all of the navigation sub-systems. This required the tying together in a common time/spatial reference of the Ferranti inertial platform, Decca Doppler and Thorn EMI SLR. This enabled such functions as Fix Monitored Azimuth waypoint storage and steering, Doppler Inertial mixing and similar modes to be implemented. As well as outputs to the navigator's displays, additional outputs went to the AFCS, moving map, HUD and, at a later date, the FLR. At this time also a conventional bombing mode was put into study.

'The critical dependence of the aircraft's operational functions on the central computing system caused Elliotts to propose a duplicated system whose concept was based in part on the fail operative philosophy of the automatic landing system then in development for the VC10. The solution chosen was to allocate "primary navigation plus secondary bombing" to No. 1 computer and "primary bombing plus secondary navigation" to No. 2. The Autonetics company was impressed with this "graceful degraded redundancy" idea and, without our knowledge, adopted it for the General Dynamics FB-111's all singing and dancing system configuration.'

Former Canberra navigator/pilot John Brownlow was serving in the Air Ministry/MoD Operational Requirements (OR) Branch as a test pilot when the TSR2 was undergoing its final development in 1963/64. 'My responsibilities included monitoring, and commenting on, the development of the TSR2 pilot's cockpit layout and flight instruments, and generally advising on piloting aspects of the aircraft. I therefore attended numerous meetings concerned with the development of the TSR2 and was able to take part in the general discussions that inevitably took place, both formally and informally, and contributed to the inevitable round of papers. The TSR2 was designed for low level operations, and therefore had a very high wing loading that would have restricted manoeuvrability at high level, and imposed limitations on cruise performance at these altitudes. The inevitable lack of flexibility that this feature imposed was often debated.

'A major concern was the reliability of the two Verdan computers that had thermionic valves. The more foresighted members of the OR Branch felt that it would not be long before a major update of the system would be required with the aircraft in service, with all the implications of cost and temporary operational limitations that this would entail. At the time, major systems, such as the HUD, navigation and automatic flight control, were being developed. The cost of these developments was being piled onto the TSR2, with much concern being expressed about the long term sustainability of these costs and, of course, the operational effectiveness of the new systems.'

Wally Mears was a Canberra navigator who was a desk officer in the TSR2 Project Office of the OR Branch. 'The OR had also been intended to give the RAF a reconnaissance capability at medium level, at least in limited war. It had been intended to operate in areas outside the NATO area that were defended by supersonic fighters, not necessarily densely defended by SAM systems. Supersonic capability was needed in such scenarios and Industry had said that a Mach 1.7 capability was "easy", given the engine thrust available. The reconnaissance community had been anxious to operate at medium level, with a sustained supersonic dash capability. The Buccaneer could not have offered such performance. The MoA's attitude had been exemplified in the early days when a supersonic variant of the Buccaneer had been offered as a solution. An unnamed Director in the MoA had written that the Buccaneer could not be made supersonic even if two Atlas missiles were strapped to it! Later, when TSR2 had been cancelled, the same Director had proposed a supersonic Buccaneer. The MoA played a game that was of no assistance either to Industry or to the Service.

'I was one of the 60 people who turned up to cockpit conferences that typified part of the control problem. Even in the MoD, there had been no centralised control of the project. There had been four staff officers in the OR office, but many other disciplines had conflicting opinions about cockpit layout. The consequence was a constant battle.

The TSR2 flight simulator. (BAe Systems Heritage)

In the end, the cockpits turned out to be reasonably well designed. The SLR, with its complex processing units, created particular problems of cockpit ergonomics. The nav/attack system, as planned, would have required a major upgrading for lack of computer capacity and because of the clumsiness of the SLR, itself dictated by the use of the nose cone for the terrain-following equipment.'

On 25 February 1965, the Deputy Chief of the Air Staff (DCAS), Air Marshal Sir Christopher Hartley, sent a note to the Air Force Board Standing Committee, which was carrying out a TSR2 Costing Exercise. It was clear and succinct:

'At the present stage of development it would only be possible to programme one mode of weapon delivery [e.g. either lay-down or dive, but not both] and ten pre-set fix points for a given sortie.'

The DCAS went on to say that all the authorities expected the demand for computer capacity to rise by at least ten per cent during development trials, and that this would mean there would be insufficient capacity for a complete operational sortie even without pre-set fix points. This situation was not that different from the limited initial operating capability of the F-35 Lighting II when it first entered service, and the universal military mantra has always been to get a new aircraft into service and to fix the bugs later. The proposed TSR2 solution was to fit Verdan DSD-1 in place of the DSD, but quite how much extra development time and cost this would have taken was not discussed.

Peter Hearne of GEC Avionics was clear that 'computer storage capacity was always a problem from Day 1. However, in early 1959 there were very few rugged computer stores of any type, and the drums and core systems of the then emerging computer industry were in no way suitable for the TSR2 flight envelope. We did have in mind to use the doubled "two-faced" disc of the marine version of Verdan [which went into the Royal Navy's four Resolution-class Polaris submarines] as an interim Product Improvement Programme for the early production aircraft. This would have given us 8,000 words total.

'Our longer term solution was based on a "soldier proof" computer which Elliotts had designed for a tank anti-aircraft artillery fire control system that had a "ruggedised" one microsecond core store – unheard of in those days – but with only 500 words of memory. This computer, suitably shrunk and micro-miniaturised, together with the other digital technology improvements of the early 1960s, became the 920M, of which some 400 went into various Jaguar aircraft. These grew in time from an 8K to a 64K store and formed the heart of an airborne computing system that drew heavily, and successfully, on the background of the TSR2.

'The Vickers/contractor integration team worked in a very harmonious fashion to produce effective systems integration. Inter-contractor relationships were undoubtedly helped by the lack of contract squabbling that almost inevitably happens on today's fixed price programmes, particularly when the price is being squeezed below a reasonable level to do the job. However, the downside was a lack of cost targets and cost control. The balance between the customer's wish (or even dream) list and his bank account was definitely not being properly monitored by the various Ministries.'

Gordon Dyer was a photo-reconnaissance Canberra navigator who was posted to the Aeroplane and Armament Experimental Establishment's Navigation Division at Boscombe Down, in Wiltshire, in January 1965 as trials officer for the TSR2 navigation system. 'In the case of the Ferranti FPS100 inertial system, it had to be surrounded by a coffin of dry-ice to keep it cool. Later models, we were told, would have had an improved cooling system. The initial trials were carried out using a Verdan computer. This was very slow and you could almost see it think when it was asked to calculate a Decca Doppler fix. The last seven flights used an Elliott Computer MCS 920, which was visibly faster. The Doppler usually performed well.

'The TSR2 flying prototypes had no components of the intended navigation fit for the operational TSR2. These flight tests were focussed on handling – i.e. engines, airframe and controls. Any navigation fit in the prototypes was basic and for safety purposes only. So apart from the thrill of the ride, any navs on those flights would not have any experience of TSR2 operational nav configuration. Cancellation of TSR2 meant we will never know whether the navigation system would have met its specification. Of course, not all was lost. Both the Doppler and the Elliott 920M re-emerged for testing and subsequent fit in the Nimrod.'

Taxi trials started on 2 September 1964. XR219 is seen here escorted by two crash tenders. (BAe Systems Heritage)

CHAPTER FIVE

FLIGHT TESTING

The first flight of the TSR2 was scheduled for January 1963, 44 months after receipt of the development contract. To meet this timeframe the companies had to begin detail design in April 1960, start manufacturing in the following July and complete final assembly in October 1962. Initially, ten, later nine, flying development plus one structural test and one fatigue test aircraft were required to obtain full airworthiness certification and to achieve release to service by January 1966.

Assembly and equipping of the first two development aircraft was to take place at Weybridge, and from there the aircraft were to be taken to nearby Wisley aerodrome, in Surrey, for the first flights. However, because of restricted runway length at Wisley, in 1963 Boscombe Down was selected for final assembly and initial flights. Final assembly of the remainder of the development batch was to take place at Weybridge and of the pre-production aircraft at Samlesbury, in Lancashire. Subsequently, in December 1964, it was announced that final assembly of production aircraft would also be done at Samlesbury and all flight testing would be conducted from Warton.

The first prototype TSR2, XR219, arrived at Boscombe Down on 6 March 1964, where it was then assembled and equipped. There was no official roll-out ceremony and XR219's removal from the hangar on its own undercarriage to an adjacent apron on 6 May was treated as a matter of course. Engine testing began two days later and was conducted continually during the summer, being interrupted only when the aircraft was returned to the hangar to be brought up to flight standard, for modifications resulting from the testing, and for systems checks.

XR219 on the flightline. (BAe Systems Heritage)

XR219 was due to fly for the first time in mid-May but teething troubles with the Olympus 22R engine led to postponements, initially to the end of June. On 24 June the Minister of Aviation, Julian Amery, advised the Minister of Defence, Peter Thorneycroft, that this flight would not take place before the end of July. It had been hoped that XR219 might make its first public appearance at the SBAC Farnborough Airshow in September 1964, but a further engine setback – fatigue failure of part of the main shaft of an Olympus during a test-bed run on 24 July – put paid to that.

The first taxi runs were made on 2 September by Roland 'Bee' Beamont, who had been appointed BAC's deputy chief test pilot, and Donald Bowen, chief test navigator. Afterwards, all taxiing trials before the first flight were made by Beamont alone. The principal object of the tests was to evaluate the performance and control of the aircraft progressively under simulated takeoff and landing conditions. This was successfully achieved, but not without encountering some minor development problems that caused delays. The main snags were hydraulic and fuel systems leaks and failure of the parachute to deploy on two occasions when the aircraft was taxiing at 140kts, which resulted in the brakes overheating and seizing. No serious defects were found with the engines and they generally performed well throughout the trials. The last taxi run before the first flight was made on 22 September and was completely successful. The next few days were spent preparing for flight and undertaking final checks.

The first flight of TSR2 XR219 was made on Sunday 27 September 1964 by Beamont and Bowen. Preparations began at 0845hrs, with engine runs lasting 45 minutes. These were followed at noon by a final taxi test to check controls and operation of the parachute and brakes, the crew having sat in their cockpits all morning waiting for the ground mist to disperse. XR219 took off at 1528hrs in clear sunny weather, escorted by a Lightning T4 flown by Jimmy Dell and a Canberra B2 piloted by John Carrodus. XR219 was airborne for 13min 50sec.

The predominant factor determining the performance of the TSR2 during its first flight was safety, it being previously decided to restrict the aircraft's takeoff weight so that flight could continue and a controlled approach and landing could be made in the event of an engine failure without exceeding the imposed engine power limitations. It was also decided to fit only that equipment essential for a first flight at low subsonic speeds so that it could be made as soon as practicable. The TSR2 therefore flew with no automatic flight control system fitted, no automatic fuel balancing, with fixed intake cones and auxiliary air intake doors and with its undercarriage extended, although the last could be retracted in an emergency. Beamont's summary of the first flight stated:

'In general the performance, stability and response to control conformed closely to the briefed values and especially to the simulator studies. Virtually all scheduled test points were achieved, and this, coupled with the high standard of systems serviceability and the adequate level of un-autostabilised control and stability in this high drag, low speed configuration, reflects a very high standard of design, preparation and inspection. In this first flight configuration and under the conditions tested, this aircraft could be flown safely by any moderately experienced pilot qualified on the Lightning or similar aircraft, and the flight development programme can therefore be said to be off to a very good start.'

Afterwards, XR219 was laid up to allow the engines to be changed and the incorporation of a number of alterations, including modifications to items Beamont had criticised in his report. Testing of the new engines started on 6 November, but this was interrupted by the need to remove them to remedy minor defects in other systems, by engine vibration trouble which necessitated replacing one of the engines, by malfunctioning of the auxiliary air intake doors and by the

A walk-round inspection before the first flight of XR219. (BAe Systems Heritage)

XR219 climbs away on its first flight from Boscombe Down on 27 September 1964. (BAe Systems Heritage)

need to check modifications to the starting system. The aircraft was again ready for flight on 23 December, but the weather was such that only a taxi test was made. The second flight was made on 31 December, lasting 13 minutes, but its scope was restricted because of a recurrence of vibration in one of the engines, which led Beamont to throttle it back to idling speed during the flight. Nevertheless, the test results confirmed the data previously collected, and showed that the aircraft's performance was the same as, or better than, predicted.

The source of engine vibration on the second flight was eventually traced to a reheat fuel pump pressure oscillation and the problem was cured before the fourth flight by fitting pumps made to higher standards of quality control. However, one problem not resolved and then receiving urgent attention was undercarriage vibration, which was felt by the crew just after touchdown on both flights as a low frequency lateral oscillation of sufficient magnitude to affect control of the aircraft on the ground. In the next phase of testing it was found that this phenomenon could be reduced by landing the aircraft at slower rates of descent, and this expedient was adopted as a temporary measure to allow flight testing to progress. A complete solution to the problem was found but TSR2 was cancelled before the necessary modifications could be incorporated.

XR219 flew for the third time on 2 January 1965. Thereafter, a sortie rate of two flights a week was maintained, which demonstrated a high degree of reliability of the aircraft's systems despite new problems with the undercarriage. The first attempt to retract the undercarriage was made during the fifth flight on 14 January, and resulted in the main bogies failing to de-rotate for the landing, which was made very gingerly. On some subsequent occasions only one of the main legs retracted completely. The undercarriage was finally retracted successfully during the tenth flight on 6 February after air-spring jacks were added to assist bogie rotation and other small modifications were made, whereupon the aircraft was taken immediately from its previous

On its first flight XR219 was limited to 240kts and an altitude of 7,000ft with the undercarriage locked down. (BAe Systems Heritage)

fastest speed of 270kts to Mach 0.8 at 2,000ft and to 511kts in a fast run over the airfield at 70ft. Thereafter, exploration of the flight envelope progressed rapidly, and by 16 February a total of 5hrs 27min flying time had been accumulated. Joining the test team were pilots Jimmy Dell and Don Knight and navigators Peter Moneypenny and Brian McCann.

On 22 February XR219 was flown to Warton for the remainder of its flight testing by Beamont and Moneypenny. They left Boscombe Down at 1315hrs and flew over Wales to Colwyn Bay, where they started the first supersonic test run northwards over the Irish Sea – a maximum speed of Mach 1.12 at 29,500ft was achieved, in which configuration the TSR2 performed perfectly. On arrival at Warton they gave a ten-minute

XR219 lands and streams at Boscombe after its first flight. (BAe Systems Heritage)

Post-first flight – Roland 'Bee' Beamont in the flying suit on the left and Don Bowen on the right. The TSR2 specification called for integral crew access ladders for use at dispersed sites. (BAe Systems Heritage)

The TSR2's undercarriage. The nose gear was a conventional twin-wheeled steerable oleo strut retracting rearwards into the fuselage. The more complex main undercarriage design was dictated by the rough strip requirement and the limited space in which to stow the landing gear. (Brooklands Museum)

demonstration of the TSR2 before assembled employees, the press and television cameras. The total flight time was 43 minutes.

In the meantime, the second prototype, XR220, had been delivered to Boscombe Down for final assembly and equipping. Unfortunately, the fuselage was accidently damaged on arrival when the lorry carrying it jack-knifed as it entered the hangar and the fuselage toppled off the trailer. Consequently, final assembly was delayed but the damage proved to be superficial. The time lost was soon regained, and by 19 February 1965 XR220 was being prepared for its first engine runs. These began five days later and were completed towards the end of March, when the aircraft was laid up for final checks before its first flight, which was planned for the beginning of April.

At the end of February, after completion of the 16th flight, which brought the total time flown with XR219 to 8hrs 11min, Beamont was able to report:

'At the end of this first stage of flight testing it can be said with certainty that TSR2 is a sound and satisfactory flying machine with superior qualities of stability and control to the predicted or better than predicted values; and there is now good reason to suggest that a high success rate may be achieved in the remainder of the CA Release programme. Without doubt the flying qualities of this aircraft are ideally suited to its design role, and it is potentially a highly successful design.'

Thereafter, XR219 made eight more sorties, all of them flown during March, which added another 5hrs 3min to its flight time. These flights expanded the low-altitude flight envelope up to Mach 0.9 at low level. They were also used to test modifications to the main undercarriage designed to eliminate the vibration present on landing. Consequently, no high-altitude flights were made and the supersonic envelope was not investigated further. The aircraft's behaviour continued to prove satisfactory throughout the trials and no major problems associated with stability, control or performance were found. Another flight planned for 2 April was aborted owing to a structural failure of the fin actuator. Before further flights could be made the TSR2 project was cancelled.

JIMMY DELL AND THE TSR2

After the first five flights of the TSR2, all made by 'Bee' Beamont, Jimmy Dell took over the main responsibility for testing the aircraft. Of the 24 flights made by the aeroplane, Dell was the pilot for half of them. An RAF fighter pilot in the early days of the jet age, he was a graduate of the Day Fighter Leaders Course at the Central Fighter Establishment, based at West Raynham, in Norfolk. An exchange tour with the USAF

ABOVE TSR2 aircrew equipment. An effective air-ventilated suit was deemed essential to minimise heating-induced fatigue at high Mach numbers. (BAe Systems Heritage)

LEFT The EE family – Lightning XR753, TSR2 XR219 and a Canberra. (BAe Systems Heritage)

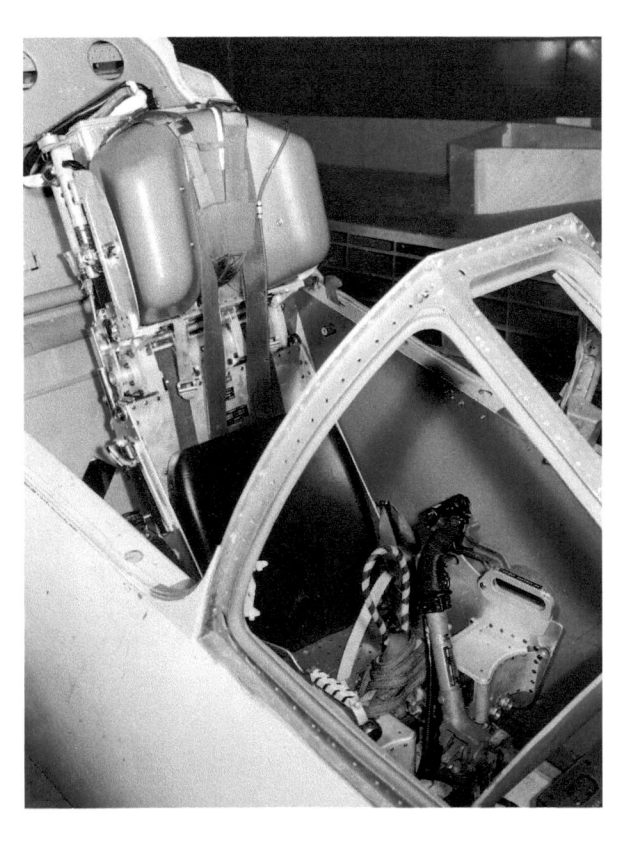

TSR2 cockpit and Martin-Baker Mk 8VA ejection seat. A single firing handle was positioned at the front of the seat base – no face blind or side handles were incorporated. It was a hard specification for Martin-Baker to get right, and the first test firing from a Meteor did not take place until 17 July 1964. The Mk 8VA seat was cancelled with TSR2, but the technology found its way into the Martin-Baker Mk 10. (BAe Systems Heritage)

and an appointment as RAF Lightning Project Pilot at Warton prepared Dell to join EE, where he became Chief Test Pilot in succession to Beamont in 1961.

'At the start of the TSR2 programme I was understudy to "Bee" Beamont, who was the designated Senior Pilot for the initial test flying. I was looking forward to many years of flying the very advanced weapons system during its development programme – which was not to be!

'The flight development programme involved nine aircraft, of which only the first flew. The second was scheduled to fly the day the cancellation was announced. The rest were in various stages of build, although it is worth mentioning that the third aircraft, XR221, had completed an initial ground run of the avionics fit, with unexpected success.

'Pre-flight preparation was standard in so much as we had the inevitable meetings, cockpit conferences, rig test experience, systems briefings and many hours in the Warton flight simulator, which, incidentally, proved very close to the in-flight experience. Having been briefed on the various systems, operating procedures and emergencies, there comes a time when in order to get the programme "off the ground", a reduced equipment and build standard is accepted that results in a re-brief on systems and equipment "as is"! The aircraft design/built standard full flight envelope of 800kts IAS [indicated air speed]/2.25 Mach/56,000ft/6.666g was reduced to take account of the actual aircraft standard for each flight.

'The aircraft I consider to be an impressive sight, and every time I visit Cosford, XR220 appears to get larger. I think the overall white finish helps. Entry to the cockpit was straightforward and the spaciousness immediately apparent. The Mk 8VA rocket-assisted ejection seat was designed to provide safe ejection at all speeds up to 650kts IAS or Mach 2.0 from sea-level to 56,000ft and, as a result of moulded seat and back panels, was unanimously considered to be the most comfortable of ejection seats by all the aircrew. The escape system provided for the usual leg restraints and also arm and head restraints. Our first experience of a torso harness was generally accepted as a step in the right direction, and significantly reduced the usual clutter of parachute and seat harness and oxygen tubes. The comfort of the ejection seat in my opinion contributed significantly to the general feeling of being "at home" with the aircraft. There was also a feeling of being "out in front" due to not being able to see the wingtips or the nose, except with the seat raised.

CHAPTER FIVE FLIGHT TESTING

'I remember on the first briefing on the escape system being told that it required 17 cartridges [to eject the crew]. On pilot initiation of ejection, the sequence was:

Immediately – Navigator's canopy jettisoned
0.2 seconds – Pilot canopy jettisoned
0.7 seconds – Navigator's seat gun fired
1.2 seconds – Pilot's seat gun fired
2.0 seconds – Crash Recorder charge fired

'On entering the cockpit the gold leaf de-misting and de-icing transparencies gave the outside world a faint golden hue which was soon forgotten until raising the canopies after flight to an often grey day. Incidentally, this triplex system was installed in Concorde.

'The cockpit layout was commendably good considering the number of personnel involved on the cockpit committee. However diligent the individuals involved, it is only in the aircraft operating environment that any deficiencies come to light. A good example was the engine throttle system, which incorporated a number of latches – from engine off to idle, max dry power to reheat and also in a reverse sense, but with an extra operation. In a dynamic situation, which required rapid reduction in power, this could, and did, result in an inadvertent shut-down of the engines, with consequent loss of some services – e.g. cooling fan operation for the wheel brakes. Obviously a modification would have been required for service clearance. An interim mod was in fact incorporated.

'Another deficiency was the miniaturised reheat nozzle and intake cone position indicators positioned at the bottom of the starboard quarter panel. An improvement in readability would have been required.

'The optical qualities of the transparencies were less than optimum, particularly the front windscreen onto which the HUD was to have

ABOVE LEFT Partial mock-up of the rear TSR2 cockpit. The FLIR display and control panel sit above the moving map display. The SLR display came next and the black circle to the left of that would eventually hold a binocular-shaped viewer for the downward sight. (BAe Systems Heritage)

ABOVE Mock-up of the TSR2's front cockpit. The moving map display is behind.

been projected. An early decision was expected to revert to a separate reflector to ensure accurate presentation of navigation and weapon aiming information. The Grumman F-14 Tomcat went through a similar process.

'Engine starting from the pilot's point of view was straightforward using the ganged rapid start bar. The actual starting required a number of complex switching actions which were controlled automatically.

'The cockpits were relatively quiet. Some initial minor problems involving the heating system afforded the pilot a warm environment but left the navigator in the cold!

'XR219 was not fitted with a HUD – the head-down standby instruments were the primary flight instruments in this case. The layout was not ideal, being on two separate planes, and the VSI [verstical speed indicator] being outside what would be considered a good scan pattern.

'Taxiing was straightforward using nosewheel steering in its fine or coarse mode. An initial feeling of over-sensitivity and a slight phase lag was quickly overcome with experience. After line-up for takeoff, the procedure was to hold the aircraft on the brakes, engage minimum reheat, check engine readings and release brakes on increasing reheat to maximum. Acceleration was impressive if not quite up to Lightning standard. Rotation initiated at 120kts resulted in a smooth unstick at 170–180kts. It is worth mentioning that during acceleration/stop tests, nosewheel lift had occurred at 105kts IAS using only ten degrees of tailplane angle, compared to a predicted speed of 130kts IAS using 18 degrees of tailplane angle. This unexpected tailplane power would possibly have resulted in achieving the OR requirement of a 650-yard takeoff distance from a semi-prepared field without resorting to the nose leg extension facility.

'Initial climb after unstick was quite steep in order to remain below the undercarriage speed limit whilst the gear went through the necessary machinations before tucking away. During the initial flights, "Bee" experienced just about all the possible parameters of undercarriage malfunction, but fortunately without the possible disastrous consequences. It was Flight 10 before we achieved normal retraction and extension.

'Once cleaned up, the immediate impression was of an exceptionally good handling aircraft and one was not conscious of the lack of autostabilisation. A criticism of control "lumpiness" found on the control test rig, and also on the aircraft during ground tests, was not evident in flight. It handled and felt like a heavy Lightning (due to higher stick forces) and there was a great temptation to treat it like a fighter and throw it around. As "Bee" had discovered earlier, the high speed low level ride qualities in the primary operational zone were outstanding due to the good gust response characteristics.

'Mild buffet was experienced in the expected areas (low speed, undercarriage and flap configurations, airbrakes), but otherwise flight remained smooth. A lasting impression was formed of the small trim

Test flight 13 – XR219 with the airbrakes open. (Brooklands Museum)

changes required during flight when making configuration changes, and even during simulated engine failure. In fact, it was possible to fly a sortie without the use of the trimmers, unless hands-off flight was required.

'At one point it became necessary to disregard the normal flight test progression and jump to designated, so called, "guarantee points". I can't remember them all, but one was Mach 0.9 at 30,000ft and another at 500kts, 2,000ft. These were carried out and confirmed predictions, so satisfying those that called for these tests. The flight envelope was also extended to 600kts at low level and Mach 1.12 at approximately 30,000ft. Handling throughout remained impressive in terms of stability and control. An early look at single-engine flying, culminating in approaches, overshoot and landing, proved successful, even with the low standard engines.

'The roll control by variable tailplane did result in a slight proverse yaw when rolling into a turn then changing to a slight adverse yaw. This was looked at on the last flight to establish the fine tuning necessary when the autostabilisation was fitted. Handling in the circuit and on the approach was eminently satisfactory, with no problem achieving an accurate and stable approach speed of 165kts. Judging the flare from an elevated position due to the nose-high attitude posed no problem, and I have recollections of a pronounced ground cushion effect that eased the aircraft onto the runway. In fact, on some flights, I was supposed to achieve something close to a no-flare landing but failed – I wouldn't have qualified for carrier landings.

'Brake parachute deployment was initiated by pulling the chute handle on the approach, which released a primary drogue (six feet in diameter). If the system was selected to manual a second pull of the handle would release the main chute. If selected to automatic the

An early TSR2 flight before trying to retract the undercarriage. The strong wingtip vortex is clearly visible. (Author's collection)

Mock-up of the TSR2's wing weapons pylons. (BAe Systems Heritage)

deployment of the main chute was operated by undercarriage switches on touch-down. The chute was 28ft in diameter, but could be streamed reefed with a diameter of 16ft by a peripheral cord. The reefing facility was designed for use in strong crosswind conditions. The two-pull facility was not cleared for the initial programme, so we used the single pull operation. During early acceleration/stop runs the brake chute system failed twice, which completely vindicated "Bee's" insistence on

When the undercarriage was eventually raised, the port main undercarriage would not retract. (BAe Systems Heritage)

a runway longer than Wisley for initial flights.

'A problem we had for the first 22 flights was that on touchdown an oscillation of the main undercarriage triggered a violent motion of the front fuselage, which threw the crew from side to side. This was for only two to three cycles but was sufficient to cause disorientation and momentary confusion on first encounter. Landings at varying rates of descent, and including landing on a foamed runway (reduced coefficient of friction), didn't provide the answer when it came to understanding the dynamics of the problem. It was eventually concluded that the frequency of the undercarriage oscillation matched that of the natural frequency of the long front fuselage, and so a fixed "de-tuning" strut was fitted to the undercarriage structure. This was assessed as successful on the last flight (with undercarriage down). The next step was to incorporate a strut as an integral part of the undercarriage to allow for retraction, but the project was cancelled before this could be achieved.

'Those of us who shared the TSR2 experience were unanimous in our conclusions that as a flying machine we had a potential world-beater.'

After one test flight the TSR2 flight observer was heard to remark, 'we flew over the sea with only one afterburner working, yet we made two chase-plane jets look like Tiger Moths'. Hyperbole apart, during its short flying career the TSR2 made many friends.

During an interview at Newark Air Museum in 2000, 'Bee' Beamont recalled that the TSR2 was 'a fantastic aeroplane to fly. We weren't able to retract the undercarriage until the tenth flight, which limited the test flying enormously – on the tenth flight we got the undercarriage away properly, did two cycles. After getting the gear to work twice,

'Bee' Beamont, the first test pilot to fly the TSR2. He was photographed here in 1957 during flight testing for the then-new Lightning jet fighter at Warton. After test flying the TSR2, 'Bee' proclaimed that 'without doubt the flying qualities of this aircraft are ideally suited to its design role and it is potentially a highly successful design'. (Photo by Popperfoto/Getty Images)

with all lights working right, I went straight out to the far extent of the test programme at that time. It had a flight resonance clearance of 500kts for that state of the flying – I took it out in stages to 500kts on that flight.

'The first time we had got the undercarriage up, it was simply superb – I was so confident in it. I ended up over Boscombe Down, where the weather was very bad – I had got Don Bowen in the back, not quite sure what was going on, with Jimmy Dell flying chase in the Lightning, trying to keep up with me in the rain and low cloud. I brought it round Boscombe's circuit thinking "This aeroplane is designed to contour fly at high speed, so let's see what it does". I brought it down Boscombe's runway at 100ft at around 450kts and the precision – it had beautiful control; I was able to relax and take my hands off the controls if I'd wanted to – was perfect. We were onto what appeared to be a magnificent technical breakthrough, which should have gone into service with the RAF in the 1970s and provided them with an aircraft that, with updating, would have been in service today. It would have had all the abilities and the modern developed equipment of the Tornado, but it would have had much further range and been a lot faster!'

Second prototype XR220 doing engine ground runs in February 1965. (BAe Systems Heritage)

CHAPTER SIX

THE POLITICS

The extent to which the TSR2 project hoovered up money was revealed by a comment from Lt-Gen Sir John Hackett, then Deputy Chief of the General Staff, in 1963:

'In view of the huge cost of the TSR2, on which the Army were bound to rely for tactical support and reconnaissance, I have serious misgivings about the extent to which it would be possible to meet the Army's needs from manned aircraft resources.'

When the British Army's Blue Water tactical nuclear surface-to-surface artillery missile was cancelled in 1962, according to Hackett, 'the current estimate of the cost of developing the TSR2 was £137m; it is now £200m'.

Such cost escalation horrified the Government. Defence Minister Peter Thorneycroft commented in a letter of 3 April 1963 that 'the cost of the present planned programme of 138 TSR2s [the same number as the British F-35 Lightning II aspiration] is of the same broad order of magnitude as the aircraft carrier replacement programme or the strategic nuclear deterrent programme', adding that this was 'a remarkable figure for a light bomber replacement'. He said that as professional advice described the TSR2's role 'as anything from a substitute for Blue Water to a substitute for a V-bomber', it seemed to him that some aspects of its role required an aircraft with capabilities of the order of those proposed for the TSR2, while others could be performed by an aircraft with capabilities 'far short of these'.

Thorneycroft also suggested that other existing or potential types should be considered 'with or without some adaptation' to discharge portions of the role intended for the TSR2 – for example, the Buccaneer, the P1154 'or any other aircraft which might be developed as a successor to the Hunter or the Lightning or the Sea Vixen'. Prime Minister, Harold Macmillan, bluntly minuted the Minister of Aviation to ask, 'Can you give me the latest position about the TSR2? What will it cost? Will it ever fly?'

A Royal Australian Air Force (RAAF) evaluation mission visited the UK to consider a possible purchase of 24 TSR2s, but then the Australian government decided to buy the TFX/F-111 – a two-seat multipurpose tactical aircraft with variable-geometry wings that was being developed by General Dynamics for the USAF and US Navy. In August 1963 the Air Minister asked the Treasury to give 'sympathetic consideration' to his Ministry's case for ordering 30 TSR2s 'to the production standard'. The Chancellor agreed to a form of words about 'a development batch order for nine and a pre-production order for 11 aircraft'. The actual announcement, made on the 30 October by the MoA was guarded:

'In addition to the orders already placed for the TSR2 . . . for development and for introductory flying by the RAF, British Aircraft Corporation have now been authorised to acquire long-dated materials to enable production of TSR2s for squadron service to begin.'

By the end of 1963 the Air Council had approved plans for the initial deployment of the TSR2 from late 1966 onwards. It was suggested that the most suitable site for the Operational Conversion Unit (OCU) – probably No. 237 OCU – where the first crews would be trained on the six pre-production aircraft, would be Coningsby, in Lincolnshire. As for low-flying training by these crews, and subsequently by crews of TSR2 squadrons in Bomber Command and RAF Germany, the most suitable area would be Libya – in conjunction with Hal Far airfield, on Malta, which would become surplus to Royal Navy requirements in 1965. The three TSR2 squadrons due to be based at Akrotiri, in Cyprus, would do their low-flying training over Turkey. Approval was given to these proposals, with the proviso that the Foreign Office should be consulted about those which concerned Libya.

The TSR2 flight test schedule projected that 'clearance of fully automatic mode [not associated with the central navigation system, bombing system and FLR] for clean aircraft should be obtained from the fourth aircraft programme early in 1966, with external stores by 1967'. Development of nodes associated with the nav/attack system were to proceed in parallel. The fourth aircraft was to support auto flight aspects of these nodes when the Boscombe programme started in 1966. David Bywater, a former Victor B1 captain who tested the V-bomber TFR, plus navigator Barry Duxbury, would have been the first serving RAF crew to fly TSR2 on these evaluation trials.

In the event, the only RAF station where a TSR2 unit did actually form was Hemswell, a former Avro Lincoln and then Thor

David Bywater, the first RAF pilot scheduled to fly the TSR2. (Author's collection)

ballistic missile base to the east of Gainsborough, in Lincolnshire. The whole TSR2 project team moved there from Weybridge on 1 October 1964 to set up the TSR2 Ground Training School. The first pilot training course on TSR2 avionics started in November 1964 and the second course was underway when the project was cancelled.

The first operational TSR2 strike unit would have been No. 40 Sqn (a Canberra unit which disbanded in 1957), and it was expected to build up to a total of 12 aircraft by the start of 1970. A second strike squadron would form within six months of No. 40 Sqn's establishment, with both being based at Marham, in Norfolk.

Notwithstanding these long-term deployment plans, the project was dogged by delays and cost overruns. On 4 December 1964 *The Times* newspaper commented that for many weeks the air had been 'thick with inflated estimates of cost on one side and exaggerated claims of performance on the other', and that 'dark rumours of cancellation' had been followed by official denials 'strenuous enough to spread panic through an arms industry'. It went on to state that the suggestion that TSR2 'was to carry the main weight of the strategic nuclear strike task between the decline of the V-bomber and the introduction of the Polaris missile' had aroused suspicions that the Air Staff 'had contrived an extension of the airborne deterrent by the simple expedient of calling it something else'. This prompted the Air Minister to say that the possible use of TSR2 in the strategic role was 'a bonus – nothing more, nothing less'.

By mid-1964 TSR2 development costs had risen to £240/260m and the production price from £2.3m to £2.8m per aircraft. In the autumn the Conservative government that had authorised the TSR2 lost the General Election. Less than three weeks after prototype XR219 made its first flight, the new Labour administration began taking a close look at the military aircraft procurement programme. The impetus was to reduce defence costs through the purchase of alternative types, including the American TFX/F-111. From 13 to 19 December a team led by the Deputy Chief of the Air Staff, visited Washington, DC, on a fact-finding mission *vis-a-vis* F-111, F-4C,

A BAC advert from 1963 extolling the TSR2's low-level capability. (Author's collection)

C-131, C-140 and Orion. A report was submitted to the Defence Minister, Denis Healey, on their return. In the meantime, the MoA had been given the task of re-examining the progress of the TSR2 and attempting to negotiate a maximum price contract for it with BAC.

As an example of the cost shavings now taking place, at a value engineering meeting on 28 October 1964, Proposal No. 44 suggested replacing the screw jacks operating the four airbrakes with hydraulic jacks. The cost saving per aircraft was put at £12,200 while the weight change was zero. This would reduce operating time from 5.5 seconds to 4 seconds, but the change would only allow emergency airbrake retraction. The Air Staff decided that emergency extension was not required – only emergency retraction was essential because remaining open could preclude returning to base.

On 2 February 1965, the Prime Minister, Harold Wilson, told the Commons that the P1154 supersonic Harrier vertical take-off fighter and the HS681 STOL transport were to be cancelled and that Lockheed C-130E Hercules and McDonnell F-4 Phantom IIs would be purchased instead. Wilson stated that development of the TSR2 would provisionally continue for performance evaluation against the F-111. He cited the probable final cost for development and production of TSR2s as £750 million, and added that a saving of

Flight 14 – the first ferry flight to Warton on 22 February 1965. (BAe Systems Heritage)

1980s TSR2

This three-view shows a TSR2 in No. 16 Sqn service in the early 1980s, painted in another low-level scheme.

some £300 million could be effected by procurement of the F-111 instead of the TSR2.

Work kept going on the 20 pre-production TSR2 aircraft and the long-lead-time items being bought. By mid-March all the major targets of the initial flight test programme had been met or exceeded but XR219 was never to fly again. The last entry in the minutes of Management meeting No. 7 held on 23 March 1965 contained the wistful line, 'A flight from Boscombe to the Paris airshow appears to be a possibility and is being explored'.

Harold Wilson said in his memoirs that there was a clear Cabinet majority for cancelling the TSR2, and instructions were given in due course for the destruction of the two completed prototypes, the five partly built aircraft on the line at Weybridge and all the remaining assemblies. On 6 April 1965, during his Budget speech, Chancellor of the Exchequer James Callaghan announced the demise of the TSR2 'forthwith'. The figure of £750 million was repeated, and it was stated that BAC and Bristol Siddeley had not been able to offer a fixed price. Options were taken out on the F-111.

The total expenditure on the TSR2 was put at £125 million. A Report of the Committee of Public Accounts for the 1966-67 Session was clear that 'on both the airframe and engine contracts the increased estimates were attributed to changes in design and specification; to successive delays; and to under-estimation of the costs and of the technical effort required'. In an attempt to learn from this abruptly terminated programme, the Committee accepted that the TSR2 was 'an aircraft of such advanced concept that it created particularly complex problems of management both for the Ministry and the contractors at a time when the latter were in process of reorganisation under the Government's plan for rationalisation of the aircraft industry. Nevertheless it seems ... that all concerned were at fault in not securing the earlier introduction of an adequate system of recording and reporting costs against physical progress to enable policy decisions to be made on the basis of up-to-date information on the financial effect of the technical problems encountered.'

In the event, the UK option to buy TFX never matured into a purchase of F-111s. Pending procurement of a more appropriate Canberra replacement, a UK frontline force of 40 free-fall Vulcans was to be kept in service, plus 16 deployed to Cyprus.

EYEWITNESS TO CANCELLATION

Sir Frank Cooper was an RAF pilot from 1941 to 1946, after which he joined the civil service and was Private Secretary to the Chief of the Air Staff before becoming Head of the Air Staff Secretariat in 1955 and Assistant Under-Secretary (Air Ministry) from 1962 to 1966. He watched the TSR2 saga unfold:

'GOR339, OR343 and the TSR2 were with us from March 1957 until April 1965. They were rarely free from controversy. To set this in context, the Empire was falling apart fast. More than 20 colonial countries became independent members of the Commonwealth or changed their status within a period of a few years. It became increasingly difficult for military aircraft to move about the world – hence the search of an "all Red Route". The enthusiasm for permanent bases, the pressure on mobility, the need for transport aircraft, the growing importance of flight refuelling and the emphasis on aircraft range and takeoff characteristics in hot climates were all by-products of this rapidly changing world. The strength of the Soviet Union, and its technical prowess, grew. The Anglo-American alliance prospered. The Sandys' White Paper of 1957, written in 11 weeks, was designed to substitute technology (particularly missiles) for people wherever possible. Conscription was abolished and there were no conscripts in the Armed Forces after 1962.

'Costs, complexity and knowledge were all escalating wildly. The functional costing system was brought in – an adaptation of the

XR219 from below. (BAe Systems Heritage)

American system that [Secretary of Defense Robert S] McNamara had introduced in the United States and subsequently urged on Thorneycroft. The MoD was reorganised. The MoS was seeking to rationalise the aircraft industry – no less than nine airframe companies responded to GOR339 in 1957. The MoA was created in November 1959 and was required to cover the whole spectrum of military and civil aviation. There were major inter-Service rows about Army helicopters, air transport, Coastal Command and aircraft carriers – all of them damaging. But perhaps above all the country was financially in dire straits for much of the time. Yet systems of budgetary control were crude and for complex projects the arrangements for forecasting, monitoring and controlling expenditure were inadequate.

'The requirement for a Canberra replacement was controversial from the start in 1957. The Royal Navy pressed, at the highest levels, that the RAF should take the NA39 [Buccaneer] and leave GOR339 to be developed in a later time-scale. These arguments were rejected by the Secretary of State for Air [George Ward] primarily on operational grounds – notably range and takeoff performance in hot climates. It is only fair to add that George Ward himself questioned how long it would take GOR339 to materialise. Exchanges between Ministers were icily polite – a forecast of what was to come. The Defence Research Policy Committee, under Sir Frederick Brundrett, was sceptical about GOR339. The Brundrett view was that NA39 could comfortably fill the tactical strike/reconnaissance role in support of the Army. The MoD, at that time a very weak department, was also sceptical and rather pro-NA39. One question it asked [prophetically] of the Air Ministry was whether the RAF had considered any foreign aircraft.

'The breach between the views of the Admiralty and the Air Ministry was complete by the summer of 1958. They went their separate ways.

The necessary approvals began to be obtained for GOR339. The hurdle of the Defence Research Policy Committee was overcome in June 1958, with DCAS [then Air Marshal Sir Geoffrey Tuttle] arguing that the requirement was vital for the three Services, vital to the aircraft industry, vital for the UK's balance of payments and to the British position in NATO. He is recorded as saying that "it would probably be the last military fighting aircraft developed in the UK". He stressed the urgency, but it was December 1958 before an unenthusiastic Minister of Defence gave his approval to the project.

'One basic problem was that, even within the Air Ministry itself, there were nagging questions about GOR339. The primary cause was lack of confidence in those concerned with research, development and production of the aircraft and the constantly changing forecasts about timing, performance and, above all, cost. From the start the overall management of the project was regarded as suspect. In some ways this was not surprising given the shotgun nature of the industrial consortium coupled with the fact that Whitehall itself spawned committees, the consequence of which was to make matters worse. To add to the confusion, the operational requirement was "upped" on several occasions.

'There was no doubt that relations with the MoS/MoA and the Air Ministry went from bad to worse, and that these poor relations spread increasingly to the MoD as a whole. The basic cause was lack of trust, particularly as regards the information received by the Air Ministry. Trust was lacking because the Procurement Ministry stood between the Air Ministry as the customer and industry as the supplier. Moreover, nothing seemed to arrive at the right time and at the right price, let alone with the desired performance.

'The lack of trust was exacerbated by the financial arrangements under which the MoS/MoA recovered production costs from the Air Ministry, but was left with the research and development costs. Hence, there was no clear objective against which the supply department could assess performance and value. The continual slippage of the forecast in-service dates added to the general air of despondency. All the change was in the wrong direction and it hurt. Slippage contributed to the ever-diminishing credibility of the MoS/MoA and of industry.

'Delays in the first flight of the TSR2 and the delay after the first flight were damaging. The engine problems – not least the blow-up of three engines – further sapped confidence. Internally, there was no doubt that the Air Ministry's own budget was over-stretched, particularly in the equipment field. The traumatic, confusing and never to be under-estimated effects of the Sandys' White Paper of 1957, which pushed for greater mobility and increased emphasis on air transport capacity, coupled with the rising cost of equipment, offer at least a partial explanation. One of the oddities was that throughout the eight-year period the need for a Canberra replacement was never seriously questioned. The need for tactical reconnaissance aircraft for

NATO, CENTO [Central Treaty Organisation, originally known as the Baghdad Pact] and SEATO was always accepted, and, despite Suez, in 1956, "East of Suez" did not become a crunch issue until after 1965, though the argument about how to project power – particularly air power – overseas ran strongly throughout. It is fascinating to recall that Harold Wilson told Parliament in December 1964 that "we cannot afford to relinquish our world role . . . sometimes called our East of Suez role".

'The Australian attitude and the Australian decision – highly costly for them as it turned out – to buy the F-111 has been the subject of considerable speculation. There is no doubt the Australian decision severely damaged the prospects of the TSR2. The MoA and the Air Ministry were open to the criticism that they could have tried harder in the early stages, though no doubt they were inhibited by the fact that there was no good story to tell, or much of a belief that the Australians would be buyers. Both Ministries tried hard before the Australians came to a final decision after the evaluation process. There was always speculation about the influence of Lord Mountbatten [then Chief of the Defence Staff] and Sir Solly Zuckerman [Chief Scientific Adviser to the Secretary of State for Defence] and their closet activities in casting doubt on the TSR2. Mountbatten actively discouraged [Air Marshal Sir Frederick] Scherger – his Australian opposite number. But talking to members of the RAAF mission, it was difficult to believe that there was much enthusiasm for the TSR2.

'The most potent factor, however, was the determination of Australia to move visibly closer to the United States in a military sense, based on its increasing doubts about Britain's military capacity to act significantly in the defence of Australia.

'What about the MoD as a whole? Successive Ministers of Defence and Secretaries of State for Air fulminated about delay, about vacillation and cost increases but stayed steady about the requirement and the means of meeting it – until the arrival of the Labour government in 1964. The Chiefs of Staff in general were supportive largely because of the need to live and let live, but also because the RAF at that time was fortunate enough to have outstanding Chiefs of the Air Staff, Vice-Chiefs, Assistant Chiefs and Directors of Plans. There is little doubt that Lord Mountbatten and Sir Solly Zuckerman would have liked to abolish the TSR2 – indeed never to have started it [a view shared by Sir Frederick Brundrett, who was Chief Scientific Adviser at the start].

'There is equally no doubt that both encouraged frequent reviews and indulged in what one might call clandestine operations. But Lord

XR219 vapour trailing. (BAe Systems Heritage)

Mountbatten was unwilling to come out into the open and tackle the issue head on, not least because it looked like disloyalty to his military colleagues [with whom he was having a tough time on other matters], though he was more than willing to encourage others privately.

'The small MoD central secretariats and scientific staffs were never enthusiastic about the TSR2 and most became hostile, but over the years most came to accept the TSR2 as water over the dam. The MoS/MoA, who in many ways were the progenitors of the project, forfeited the trust of all primarily because of the inaccuracy of their forecasts and their inability to oversee and organise the management of the project as a whole. The Treasury sought every opportunity to express doubts and misgivings about costs and to encourage review of the project and its management. It was worried that by some means or other, which it could not clearly discern, the role of the TSR2 was being extended and the specification increased upwards. Basically, however, the Treasury accepted the need for a Canberra replacement, and it would be difficult to argue that it seriously held the project up on financial or other grounds.

'What is surprising is how ineffective were those who believed wholly or partly that the RAF should adopt Buccaneers. Some of the critics were in positions of great responsibility but were incapable or unwilling to exercise it.

'Major defence changes stem in the main either from external influences or from economic and financial factors. It was a combination of these that broke the TSR2's back. The cost history of the TSR2 was horrific. It is worth remembering that the Air Ministry's total budget averaged little more than £500 million a year over the aircraft's eight-year development period, while that of the MoA during the same period was £200 million. The first figure informally bandied around for GOR339 was around £16 million – that was a guesstimate. In July 1958 the MoS told the Treasury that the in-house estimate was £35 million to CA Release. In November 1959 it was £62 million up to CA Release, with plus possibly £15–25 million on top. In 1960 there was much discussion in Whitehall about astronomic costs. Confidence was beginning to sag.

'There is no doubt that during the first half of 1964 there was a change of attitude in the Air Ministry among some senior people [including the CAS, the Director of Plans (Air) and the Secretariat], and the Air Ministry began to question seriously whether the RAF programme could bear the cost of the TSR2, and about its effectiveness

---

NEXT PAGES

TSR2 in Operation *Granby*

The RAF's primary strike aircraft during the campaign to evict Saddam Hussein's forces from Kuwait, the TSR2 was by this time equipped with a modified anti-runway munitions system, JP233, fitted into its bomb-bay. This is one of No. 617 Sqn's TSR2s in the 'desert pink' colour scheme used by RAF units involved in the Gulf War.

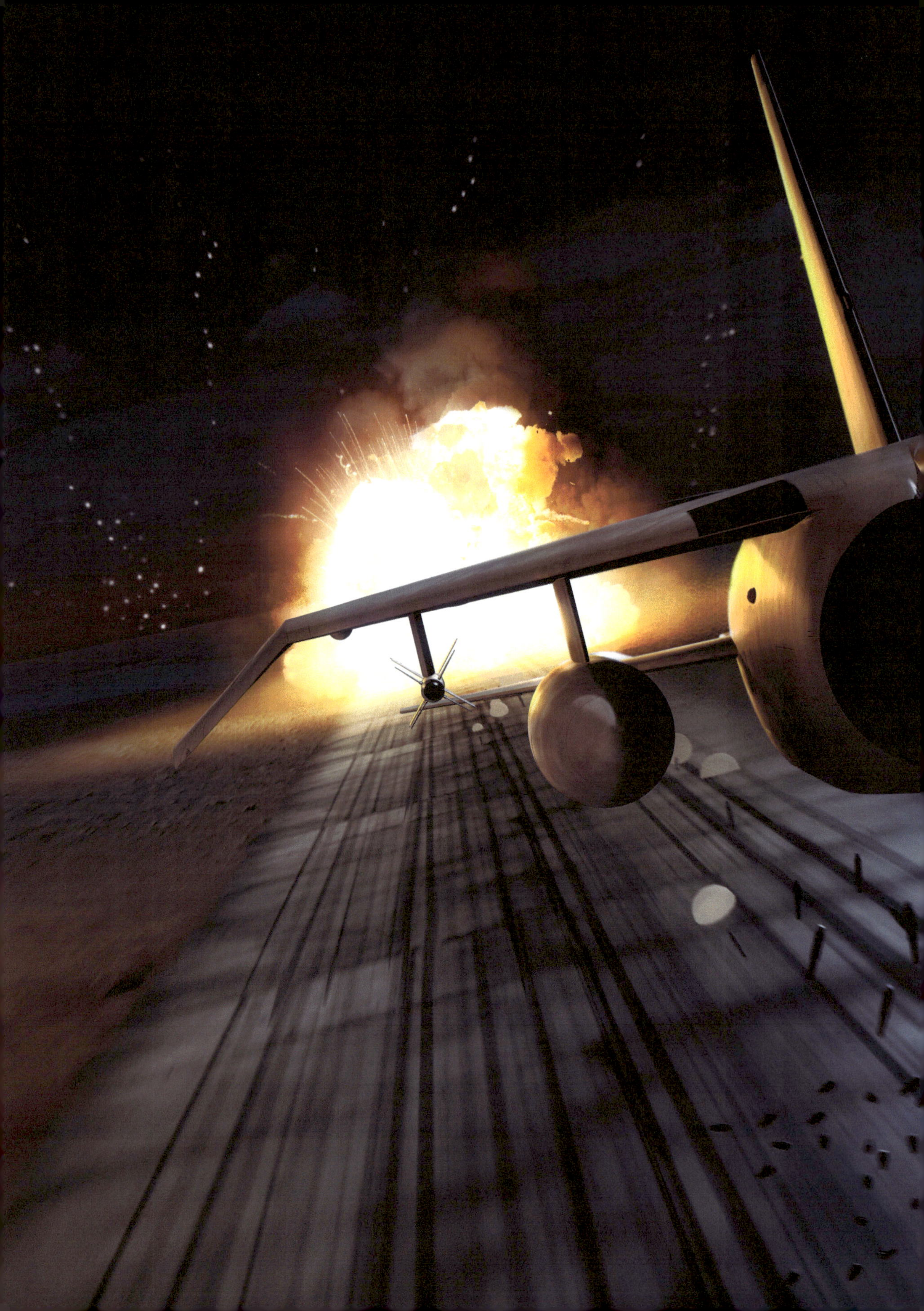

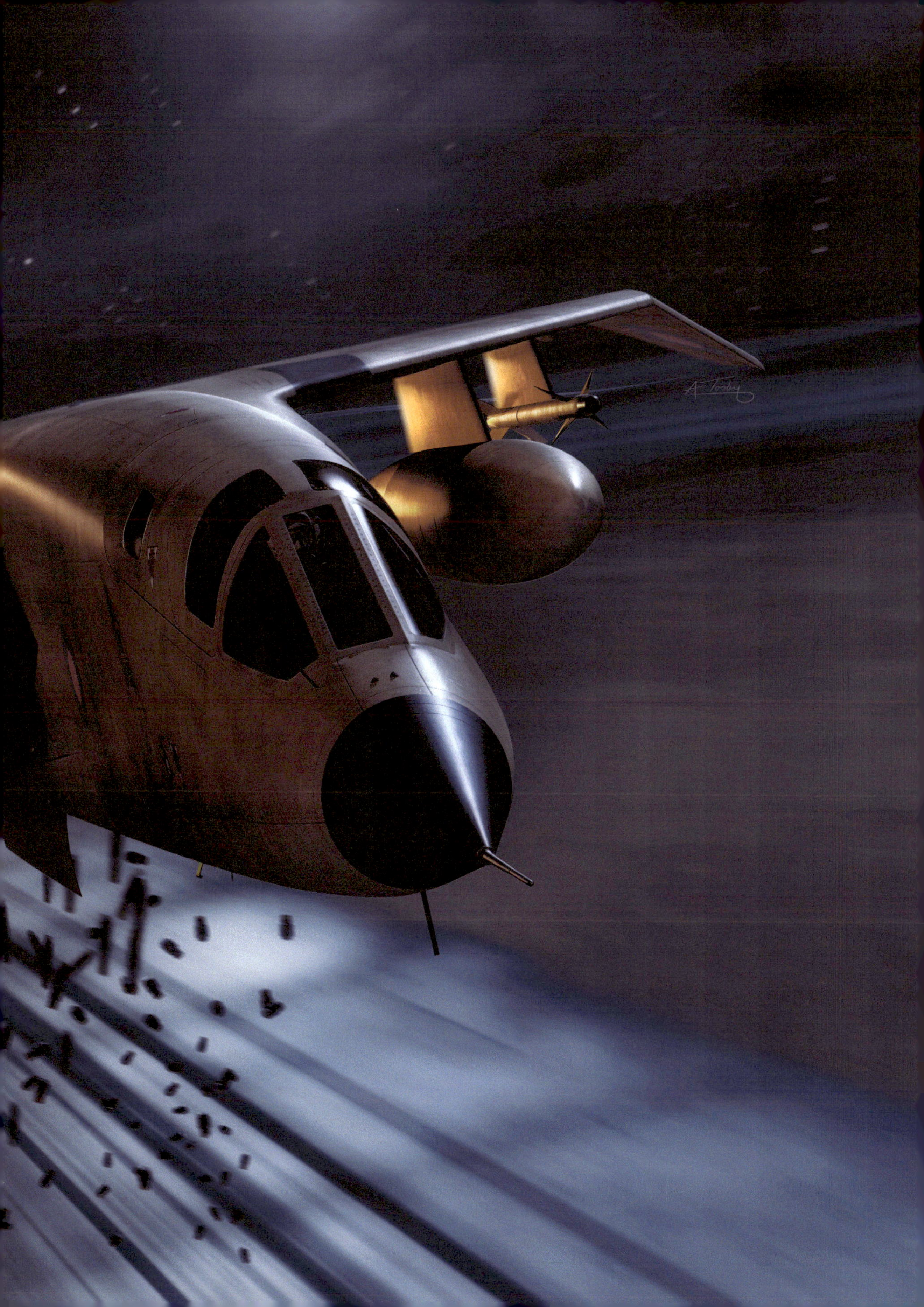

in terms of performance. One consequence was that the Air Ministry decided to ask the MoA in March 1964 for a fixed price contract. Another was that the CAS [Air Chief Marshal Sir Charles Elworthy], after discussion with a very limited circle, took a note from himself to the Secretary of State for Air [Hugh Fraser], who showed it to Thorneycroft expressing doubts about the project. The paper was torn up and the CAS told that this was not a matter to be discussed before a general election.

'In truth, by this time most of the Air Ministry had lost confidence in almost anything to do with the TSR2. In January 1965 the Defence and Overseas Policy Committee virtually wrote the death knell of the TSR2 because "US aircraft were available at fixed prices and with fixed delivery dates, they cost much less and would be in operation earlier". What had almost certainly been underestimated was the impact of the favourable financial terms skilfully presented by the Americans, and which included a potentially favourable impact on the RAF programme as a whole. By the time of the October 1964 general election the cost of the TSR2 had gone through the roof. The Americans had put forward proposals for offsets covering the three American aircraft [the F-111, the Phantom II and the Hercules], plus a deferred payment scheme, to be financed by the US Export–Import Bank. The bill was to be met by 14 half-yearly instalments, with interest at 5¾ per cent. The consequence would be to flatten out the hump in Air Ministry expenditure and defer much expenditure for several years.

'It is extraordinary how little space the cancellation of the TSR2 takes in Harold Wilson's autobiography. In Denis Healey's autobiography he states that the 1960 estimate had tripled four years later to a sum of £750 million, the delivery date instead of being 1965 had slipped to 1968 or 1969, and £250 million, he claimed, would have been saved over ten years by buying the F-111, which was itself axed from the programme in 1968. Roy Jenkins [Home Secretary at the time of the TSR2's cancellation] was also brief, but it is perhaps worth quoting what he said:

'"The Australian Air Force had in 1964 delivered a nearly final blow to the TSR2 by opting for the F-111. By early 1965 the British MoD, Air Marshals as much as Ministers, wanted to do the same. The TSR2, good plane though it was, had few friends outside the aircraft industry and the military chauvinist political lobby. I did not think we should keep it going, although I was not convinced that the automatic alternative was to buy the F-111. My scepticism about a continuing British "East of Suez" role predisposed me in favour of doing without either. This divided me from Healey, who was determined to buy the American plane.

XR219 with bomb-bay doors open. (Brooklands Museum)

'"The Treasury were naturally in favour of saving money, although its voice was rendered uncertain by Callaghan being as an instinctive "East of Suez" man as I was a sceptic. But he certainly wanted the TSR2 axed. An extraordinary and complex story. Cancellation was inevitable."'

Air Chief Marshal Sir Patrick Hine, fighter pilot and commander of all British Forces during Operation *Granby* ('Gulf War I') in 1991, was a student at the RAF Advanced Staff College when the cancellation of TSR2 was announced. 'At the end of that year I was posted as Personal Air Secretary to the Minister of Defence for the RAF – Lord Shackleton. As Sir Frank Cooper said, the general view in the MoD at the time, including that of the then CAS, was that cancellation had been inevitable – on the grounds of unaffordability and with rising costs that were out of control.'

Sir George Edwards of Vickers had been much occupied with the setting up of BAC, but near the end of 1964 he put Freddie Page in charge of the TSR2 programme. In Freddie's words, 'This was a doubtful honour as it was by then clear that a new Labour government, aided by several powerful figures behind the scenes in Whitehall, was determined to cancel the project. When the project was cancelled, many thousands of people in industry lost their jobs, but I am not aware that the politicians, civil servants and Service personnel responsible for over-specification and lack of control suffered equally. Finally, it is my firm belief that, if the original P17A proposals had been accepted, the TSR2 to a very advanced standard might well still be in service with the RAF today, and with several other air forces as well at considerable profit to the UK.'

The epitaph for TSR2 was best put by Sir Sydney Camm, designer of the Hawker Hurricane. 'All modern aircraft have four dimensions: span, length, height and politics. TSR2 got just the first three right.'

The bitter end. The TSR2 mock-up being burnt at Warton on 30 June 1965. (BAe Systems Heritage)

CHAPTER SEVEN

WHAT MIGHT HAVE BEEN

XR219 being welcomed at Warton on 22 February 1965. (BAe Systems Heritage)

In 1957 EE envisaged a Mach 1.6 twin-engined aircraft weighing less than 70,000lb, while Vickers proposed a less than 50,000lb single-engined aircraft. By 1965 the TSR2 had ended up as a 100,000lb+ twin jet. Although the required sea level speed, takeoff distance and radius of action remained roughly the same, the altitude speed for design and test had increased from no specific figure to Mach 2.25 and 825kts IAS at maximum worldwide temperature. Low-level penetration height had dropped from 1,000/1,500ft to not more than 200ft and the runway classification number decreased from 40 to 22, with a tyre pressure less than 80psi. At a late stage, the main computer installation was doubled. At one meeting with officials when the design speeds and temperatures were increased, EE Chief Executive Freddie Page said, 'Gentlemen, I hope you realise that what you have done will ensure that this project will cost the earth'.

A 1959 document listed the following conventional attack targets envisaged for the TSR2 – bridges, blast-resistant and reinforced-concrete buildings, radio stations, airfield runways, parked aircraft, thin-skinned and armoured vehicles, small ships, guided-missile sites and even 'tribal forts' outside Europe. However, as its name implied, the TSR2's main role was Tactical Strike. Although the aeroplane would have been able to carry conventional iron bombs plus at least two Anglo-French Martel missiles (both TV-guided and anti-radar versions) on underwing pylons, the TSR2's primary payload was to be the standard RAF tactical 'nuke'. This was Red Beard in the early 1960s, and 142 of these 10–15KT nuclear weapons were acquired between October 1960 and May 1963, 48 of which were to be stored in Tengah, Singapore, and 32 in Akrotiri, Cyprus.

Four squadrons of Canberras based in West Germany were assigned to Supreme Allied Commander Europe (SACEUR) in the same way as the three squadrons of Valiants based at Marham. In 1962 the CAS, Marshal of the RAF Sir Thomas Pike, stated that 'TSR2 was essential to the RAF as a general workhorse in replacement of the Canberra, to provide close support for the army with atomic or conventional bombs and for reconnaissance. It was required to meet NATO, CENTO and SEATO commitments . . . The plan was to deploy roughly one for every two Canberras, a total of 106 in all.' By then the TSR2 was also seen as a likely replacement for the Valiants assigned to SACEUR.

In August 1959 first drafts of an Air Ministry requirement for an improved kiloton bomb (OR.1177) and its warhead (OR.1176) were circulated. The Ministry's overriding concern was to acquire a weapon suitable for use on the TSR2, mainly against hard targets from low level at supersonic speed. OR.1177 was issued formally as a joint naval and air staff requirement at the end of May 1960, with the new weapon required in service by 1964–65. A bomb weight of 900lb and diameter of 20in were envisaged, and the warhead would need to be variable in yield from 10–300KT.

A retarded or lay-down mode, to allow accurate delivery from low level without the need to 'pop up' into enemy radar cover, was an important part of the OR.1177 requirement. Parachute retardation of the falling bomb would reduce its speed to 40ft/sec in the case of a 'fragile' warhead, or 250ft/sec in the case of a specially 'ruggedised' warhead (the bomb had to be tough enough to survive the initial impact without destroying the fusing system and disrupting the structure of the 'physics package' that made up the warhead). As an insurance

PLANNED TSR2 FLIGHT PROFILES						
Profile	Fuel load	Altitude	Speed	Distance	Still air time	Remarks
Economic cruise	Max internal	23–35,000ft	Mach 0.92	2,780 miles (4,470 km)	5hrs 5min	Ranges based on 2,000lb weapon load carried internally. Normal fuel reserves included
Economic cruise	Max internal plus 2 x 450 gal wing tanks plus 1,000 gal ventral tank	15–35,000ft	Mach 0.88–0.92	3,440 miles (5,540 km)	6hrs 20min – 6hrs 35min	–
Low-level cruise	Max internal	200ft above ground level	Mach 0.90	1,818 miles (2,930 km)	2hrs 40min	–
Low-level cruise	Max internal and 2 x 450 gal wing tanks plus 1,000 gal ventral tank	200ft above ground level	Mach 0.90	2,060 miles (3,320 km)	3hrs 30min	–
Supersonic cruise	Max internal	50–58,000ft	Mach 2.00	1,000 miles (1,600 km)	53min	Climb/descent at less than Mach 2.00 (airframe and engines stress limited to 45min at Mach 2.00)

against cancellation of the Douglas GAM-87 Skybolt air-launched ballistic missile, pressure grew for a TSR2 lay-down weapon not only with a kiloton yield for tactical use, but also in a megaton strategic version. By December 1961, the MoA reference number WE177 had been allocated to what would become a family of tactical bombs.

When a new version of the OR.1176 requirement was circulated in November 1962, it stated that 'The tactical targets envisaged by the Air Ministry include airfields, missile sites and general communications in the middle and far eastern theatres. Studies show that most of these targets could best be attacked with weapons in the 50 to 300KT range, allowing for degraded delivery accuracy in the face of defences or weather hazards . . .'

The WE177 was to be capable of being freefall delivered from high level by the TSR2, dive/toss from medium level by the TSR2 and laydown from as low as 50ft by the TSR2. Carriage of two WE177s internally in the TSR2 implied a maximum rear fin diameter of 24in – the Air Ministry was keen to avoid any specification changes to the TSR2's weapons bay.

The original low-yield version became known as the WE177A, although the high-yield WE177B was given higher priority and would enter service sooner. Out of 102 warheads for lay-down bombs approved for the TSR2, around half were to be of the high-yield WE177B version – 'B' for the 'Big One'.

A standard 1,000nm TSR2 mission was to be flown with a 2,000lb internal weapon load (the weight of a Red Beard) and a reserve of five per cent of takeoff fuel, plus eight minutes loiter. Of this radius, 100nm was to be flown at an altitude at Mach 1.7 while 200nm into and out of the target area was to be flown at Mach 0.9 at 200ft. The remainder of the mission was at Mach 0.92 at altitude. A lo-lo sortie at Mach 0.9 at 200ft gave a radius of action of 700nm. Extra fuel was available in the form of 450-gallon jettisonable underwing tanks, a 570-gallon tank to be carried in place of bombs in the weapons bay and a jettisonable ventral tank of 1,000 gallons under the fuselage.

Before takeoff, the sortie flight plan was to be fed into the Verdan digital computer on punched tape. Because of the long-range nature of TSR2 sorties, the navigator was to obtain an independent fix every

The TSR2 bomb-bay. (BAe Systems Heritage)

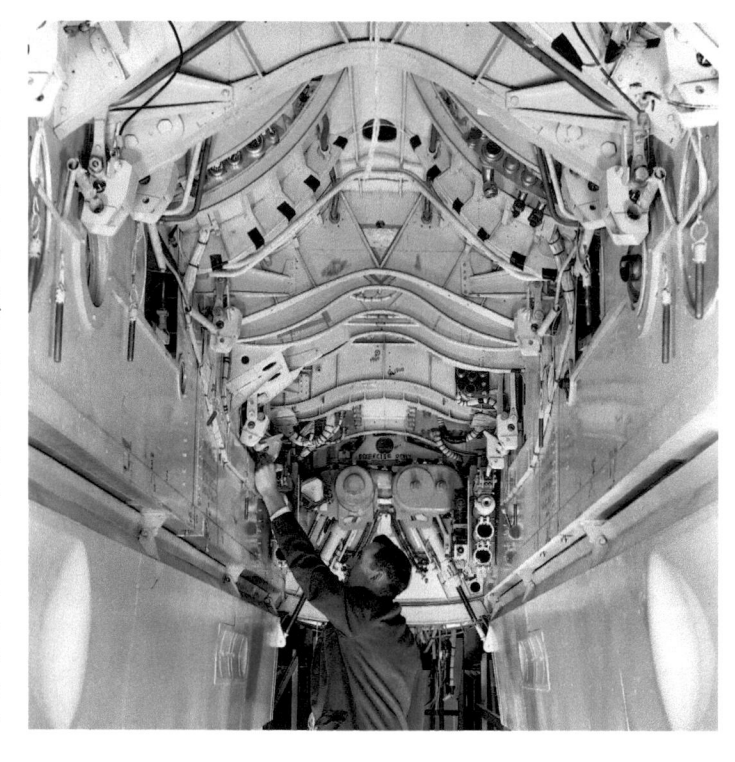

ABOVE The TSR2 bomb-bay with two dummy WE177 nuclear weapons uploaded. Fitting them in parallel released space for a forward bomb-bay fuel tank. (BAe Systems Heritage)

ABOVE RIGHT A pair of dummy WE177 nuclear weapon shapes in line astern. In reality, only one of the large-yield WE177B bombs would have been carried. (BAe Systems Heritage)

100 miles or so over well-mapped terrain. The computed fix on the radar display was synchronised with the actual fix painted by the SLR. The final element in the overall nav/attack system was terrain-following for continuous flying at a height of 200ft. The data came from the Ferranti monopulse FLR, which gave the computer information on the range and angle of the terrain ahead. The resulting signals were fed to both the AFCS and the HUD for either automatic or manual flight. The terrain-following system was likened to a stiff spring extending forward of the aircraft and slightly convex on the side touching the ground. The spring riding over the undulations in the terrain raised or lowered the flight vector as required. This was known as the 'ski-toe' locus. The TSR2 nav/attack and control system was very ambitious for the mid-1960s, and it was a forerunner of the 'gee-whizz' aids that are now essential on modern fast jet combat aircraft. Much of the TSR2's avionics technology ultimately found its way into what became the Tornado GR1.

But the TSR2 was not stealthy. The airframe had been designed to minimise radar detection from head-on, particularly in the X-band, and as a result flat, forward-facing bulkheads were kept to a minimum and the engine compressors were partly hidden from direct forward view by long, twisted intake tunnels. The intake tunnels, while masking the hearty Olympus engines, were nonetheless a source of considerable reflection, and the thermal and structural loads to which they were subject precluded the application of radar-absorbing material (RAM).

These intakes provided approximately 60 per cent of the total frontal radar cross-section of 20m². In March 1960 the Air Staff agreed that no RAM should be applied to the intakes. It was suggested that research should begin on the creation of a suitable high-temperature and high-strength RAM for future use on other aircraft, and for application to the TSR2 as an in-service upgrade.

An early EE study looked at the problem of vulnerability versus altitude and whether or not radio countermeasures (RCM) and electronic countermeasures (ECM) were worth carrying on the TSR2. Two anti-aircraft systems were examined – the American MIM-23 Hawk and the British Yellow Temple (the radar for the EE Thunderbird SAM), both of which were specifically designed for defence against low-level bombers. The study concluded that, although contour flying could delay the moment when the first weapon could be fired, space should be made available in every aircraft for the installation of an X-band carcinotron (or noise generator to prevent missile lock-on), plus decoy rockets, if the aircraft was to routinely attack heavily defended areas.

However, cost-overrun pressures led to passive and active RCM/ECM equipment being 'salami-sliced' from the TSR2 requirement. Naval/Air Staff Targets 830 (jammer), 836 (towed decoy), 837 (combined pod to include jammer, towed decoy, chaff and flares) and 841 (draft for a passive-warning IR detector) were all being considered at the time of its cancellation. The emphasis lay in getting the TSR2 into service with an Initial Operating Capability. Funding for refinements such as ECM would be found later.

A February 1963 RAE paper suggested the need for a stand-off weapon that was light enough to be carried by the Buccaneer, or at least the TSR2, but with enough range – between 25 and 60 miles – to avoid the need to penetrate the majority of Soviet city defences. But the TSR2 was a 'tactical' strike aircraft that was never procured to attack Soviet cities – that was the job of Blue Steel-equipped V-bombers and then the Polaris-equipped nuclear submarines. That did not stop Bristol from proposing a new family of cruise-type missiles for the TSR2, and there was even talk of a WE177 mounted on the front of an unguided rocket. But with the TSR2 costing as much as it did, there was never to be any spare funding for missile add-ons. If the TSR2 had entered service, it would have used the nuclear, conventional and airfield denial weapons procured for RAF Tornados. The same would have been true of ECM pods.

In strike mode the TSR2 was also required to carry out the photographic and/or radar reconnaissance roles. Strike/reconnaissance

The TSR2 bomb-bay with 6 x 1,000lb dummy bombs installed. (BAe Systems Heritage)

sorties were performed using the aircraft's basic equipment, comprising the SLR and three cameras permanently installed in the nose fuselage. In the dedicated reconnaissance fit, the TSR2 was to obtain tactical information to support both the counter-air and the land battle. The reconnaissance-optimised TSR2 had the navigator's bombing control panel replaced by a reconnaissance control panel and a large reconnaissance pack mounted in semi-conformal style below the fuselage. This reconnaissance pack contained three separate systems – optical Linescan, SLR and photographic cameras. A mock-up pack was fitted to XR225 at Weybridge in February 1965, and this highlighted the need for some pack and airframe modifications. These were to be incorporated on airframes from XS665 onwards.

Eight TSR2s, fitted out for reconnaissance, were to have formed a third squadron after the two strike squadrons had been established at Marham. The Director of Air Staff Plans, Air Commodore F Desmond Hughes, wrote to the Commander-in-Chief Bomber Command, Air Marshal Sir John Grandy, in April 1964, stating that re-equipment of a Canberra PR7 squadron should be completed in early 1969. However, as with the RAF Tornado force 30 years later, the nature of the TSR2 tactical reconnaissance fit would alter, with advanced podded equipment replacing the original sensors. And then there would no longer be any need for a specialised cadre of reconnaissance crews.

TSR2 armament loads. Future loads such as electronic warfare jammers, decoys and flares would have been carried on the lighter, outer underwing pylons. Heavier loads such as drop tanks and JP233 airfield denial weapons would have gone on the inboard. (Brooklands Museum)

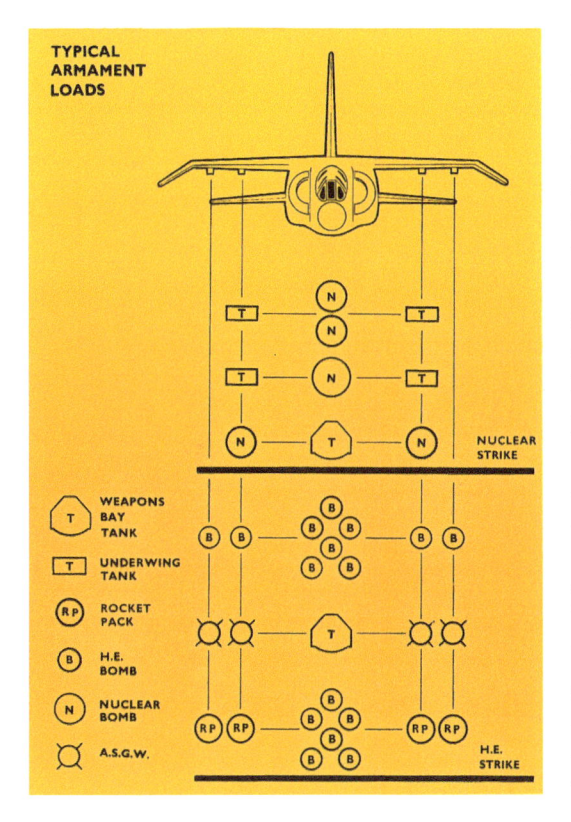

TSR2 – AN OPERATIONAL SUCCESS?

How do those who would have operated TSR2 think it would have fared? Tom Eeles went through the RAF College at Cranwell from 1960 to 1963, and subsequently joined No. 16 Sqn in RAF Germany flying the Canberra B(I)8. He then went on to fly the Buccaneer:

'With the benefit of hindsight, the TSR2 hardly seemed to be a "tactical" aircraft, but more a strategic asset to replace the V-bombers. It was very fast, but looking at the wing, probably with a poor turning performance and very prone to inertia coupling [long, heavy fuselage/tiny light wings]. So probably it would have been good for singleton nuclear strike missions but not as an agile aircraft.

'How many aircraft were we to have? Probably about 50–70 at most given the cost, so I suspect basing would be in the UK rather than Germany. Operations from rough strips? Highly unlikely given the degree of support needed. The arrival of much larger numbers of genuinely tactical aircraft – Phantom II/Buccaneer/Harrier/Jaguar – changed the RAF into a genuinely tactical air force, a process

CHAPTER SEVEN **WHAT MIGHT HAVE BEEN**

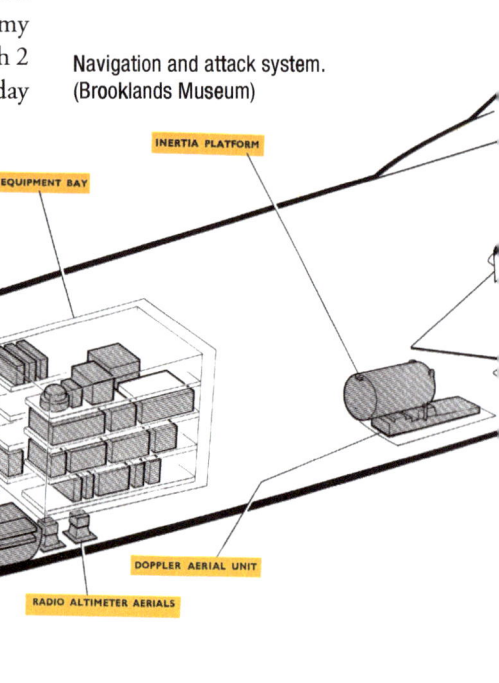

accelerated with the moving of the deterrent to the Royal Navy.'

Kevan Dearman served as a pilot with No. 6 Sqn, flying Canberras from Akrotiri:

'In my logbook I was certified to fly over Turkey at 50ft. Canberras in those days ringed the USSR. I was going to go onto the TSR2 and I recall going into the secure target study room with the TSR2 handbook. When it was cancelled I was going onto the F-111, and when that was cancelled in turn, I ended up instructing on Chipmunks with the Oxford University Air Squadron. The TSR2 needed its speed to take on the Soviet defences, but getting it into service would have bankrupted the RAF.'

Test pilot John Brownlow was close to the TSR2 design process during 1963 and 1964:

'At the time, I believed the TSR2 concept was correct, and that the TSR2 would have done the job specified in the OR [fast, low level, conventional and nuclear using UK and overseas bases]. However, other more flexible aircraft designs were in the pipeline, and so it has turned out. I do wonder whether we could have made the very complex TSR2 work. In true RAF fashion we would probably have coped!'

Jock Heron flew Hunters with Nos. 43 and 54 Sqns, before joining the Air Fighting Development Squadron at the Central Fighter Establishment (CFE). During his time at the CFE he flew the Lightning and the Mirage III, before undertaking an exchange tour with the USAF at Nellis, flying the F-105 Thunderchief, which made him qualified on Mach 2 aircraft from three nations:

'My tour at Nellis, I suspect, would have been a step towards involvement with TSR2 because I had served a tour at CFE as a trials pilot, and my return to the UK in 1967 would have coincided with the formation of a trials unit. There is much to debate about the TSR2. Strong views are held on both sides, but where I thought that my predecessors in OR were being over ambitious was aiming for Mach 2 from a grass field! We learned our lesson, which is why Tornado today is a much better option, even though we didn't get that all right. If they'd updated the TSR2 engines to turbofans and digitised the avionics, the TSR2 might still be in service today. Mach 2 was too ambitious, but I can understand the

Navigation and attack system. (Brooklands Museum)

FORWARD LOOKING RADAR
SIDEWAYS LOOKING CAMERAS
EQUIPMENT BAY
U.H.F. AERIAL
RADIO ALTIMETER AERIALS
FORWARD/DOWNWARD LOOKING CAMERA
DOPPLER
RECONNAISSANCE PACK

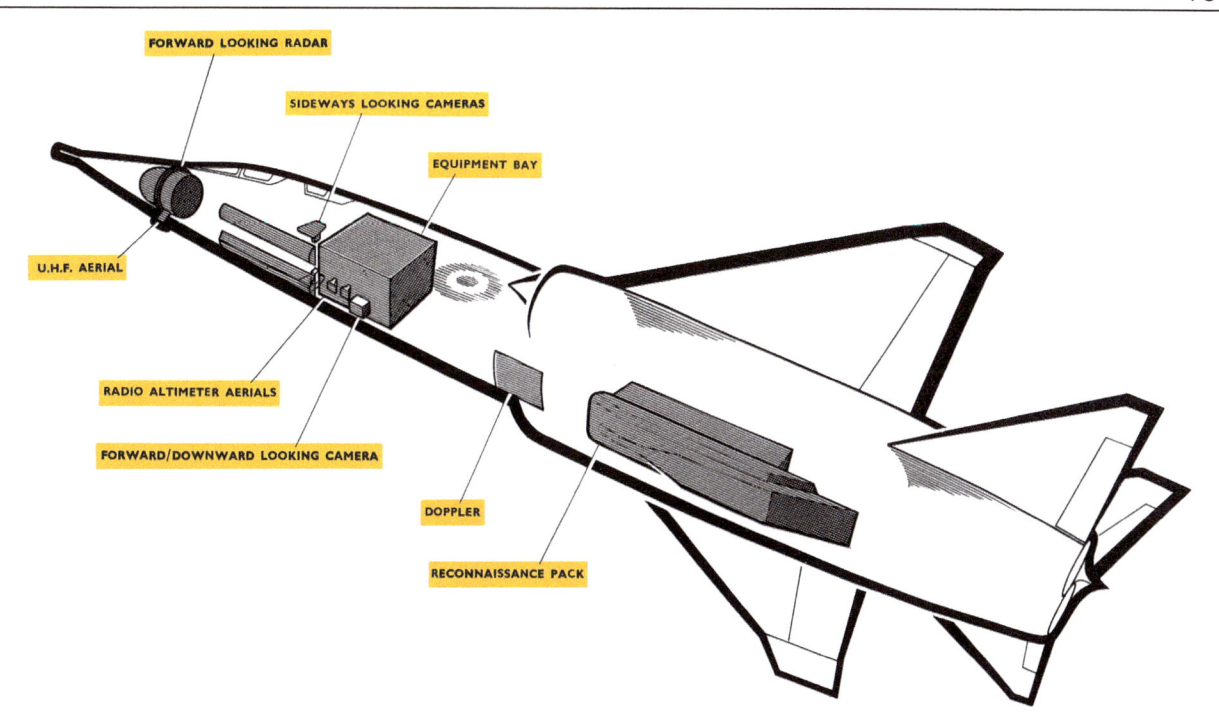

Underside view of the TSR2 reconnaissance installation. (Brooklands Museum)

thinking at the time.'

Graham Pitchfork was a Canberra PR7 navigator who did an exchange tour on Royal Navy Buccaneers to prepare him for being among the first TSR2 crews:

'I wonder if the two conflicting OR requirements of 600yd/LCN 20 [runway classification number] takeoff performance and Mach 2.25 at high level were ever reconcilable.'

Jimmy Dell was always puzzled with the requirement for 650kts out of a semi-prepared strip:

'Who, I ask, would have flown the aircraft into such a strip in the first place, let alone providing the sort of logistic support necessary for such a sophisticated aircraft? I still have a copy of the Flight Manual for the TSR2, and when I show it to modern Tornado pilots their reaction is to say that the TSR2 was just a longer range Tornado. The Tornado flight envelope (800kts/Mach 2.2) is nearly identical, yet the TSR2 was cancelled more than 30 years ago.'

Higher up the command chain, Operation *Granby* commander Sir Paddy Hine has a view of how things would look now if the TSR2

NEXT PAGES: Operation *Allied Force*

By the 1990s, the TSR2 had been updated with a retractable air refuelling probe, modern avionics, and could carry ECM pods, chaff dispensers and modern precision guided weapons. It was, however, ageing, and was due to be replaced by the stealthy Anglo-French-German Europanther. Here, TSR2s of No. 31 Sqn are shown in action during 1999's Operation *Allied Force*. Armed with 2,000lb Paveway III bunker-busting bombs, which were precision-guided by TIALD laser designators on the forward fuselage, their mission is to destroy bridges and tunnels on the main supply route between Serbia and Kosovo.

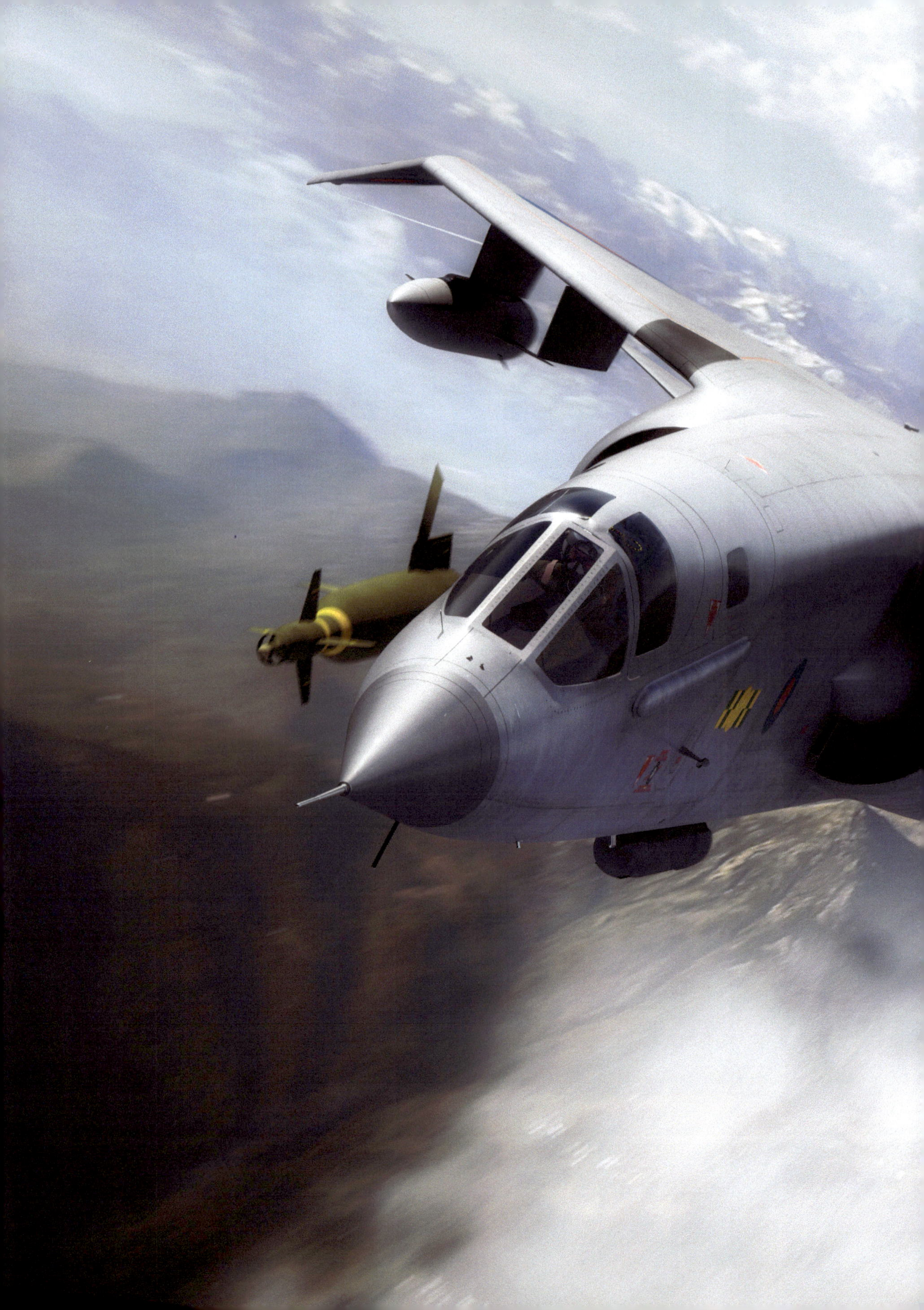

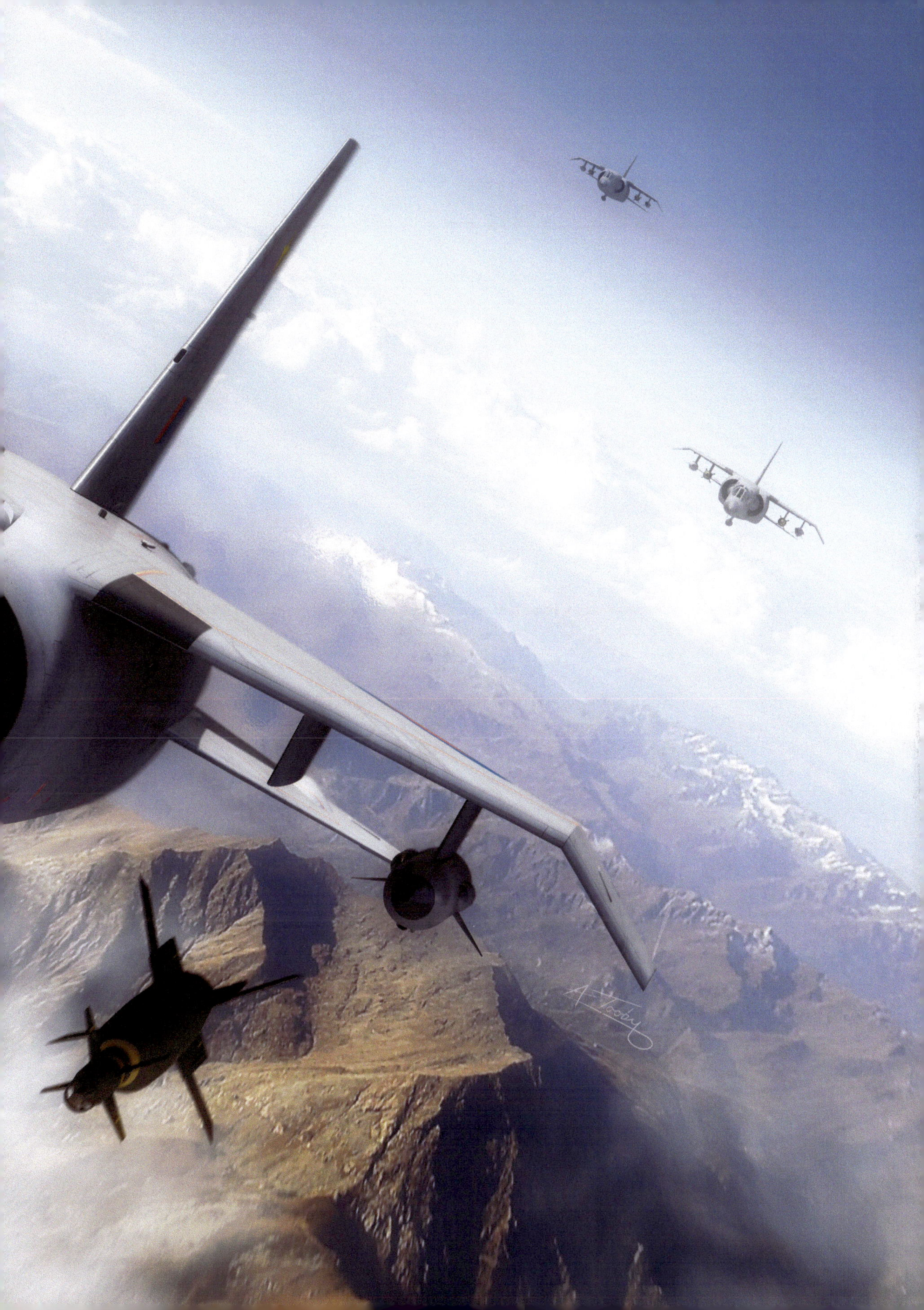

CHAPTER SEVEN WHAT MIGHT HAVE BEEN

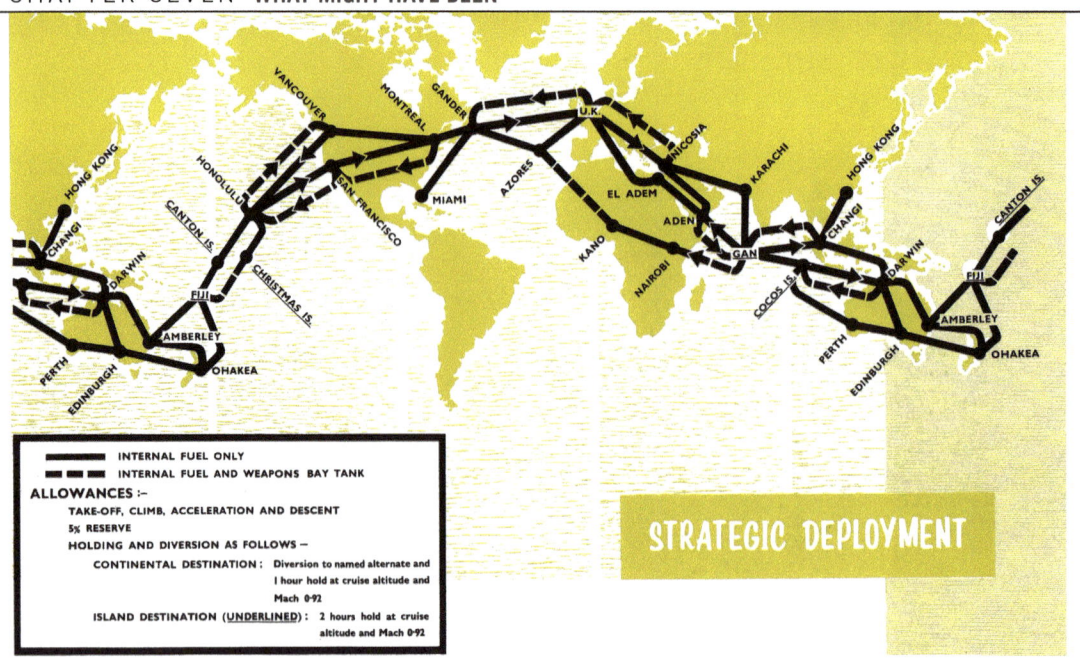

The TSR2's strategic deployment capabilities, with and without the bomb-bay fuel tank. The TSR2 could have had enhanced worldwide ferry capabilities with underwing drop tanks. (Brooklands Museum)

had survived:

'To do so I must make at least one assumption, and that is that while the TSR2 project proceeded, the earlier cancellation of the P1154 stood. It is very important that I put that peg in the ground because I believe it was right to cancel the P1154. The highly effective off-base operating capability developed by the RAF with the Harrier could not have been achieved with the P1154 with its plenum-chamber burning reheat system, which would have caused very severe ground erosion problems. In short, we would have been trying to run before we could walk, and that could have had a most adverse, if not fatal, impact on V/STOL in the RAF.

'Here then is my TSR2 scenario. First, the RAF would have got about a decade earlier the kind of capability it eventually enjoyed with the Tornado GR1. The avionics may not have been quite so well advanced, nor would the TSR2 have been so manoeuvrable, but it would have had longer legs. The bottom line is that the TSR2 showed all the signs of being a better aircraft than its nearest competitor, the F-111, but there remains a big question mark over cost and, therefore, ultimate affordability and cost-effectiveness.

'Next, let us have a look at force structure. TSR2 would have replaced the Canberra but, because of high costs, not on a one-for-one basis. Probably no more than 100 aircraft would have been procured. The RAF's strike/attack/recce force would thus have become smaller, unless a second aircraft had been procured, for which money would almost certainly not have been available.

'By 1965 the days of the V-bomber force were numbered as a result of the Polaris decision. The Air Staff would probably have argued for more TSR2s to replace some of the V-bombers – but only once the programme was secure – and they may not have been successful. TSR2 would have been used for nuclear strike/deeper recce, Offensive Counter

Air and interdiction, but not, except in extremes, for Battlefield Air Interdiction and Close Air Support [CAS]. It was not tailored for those missions and it would not have been cost-effective. Therefore, another aircraft would have been needed to replace the Hunters in the UK, Germany, the Gulf and Far East.

'Would this second aircraft have been the Harrier or Jaguar, or perhaps a multi-role fighter like the STOL F-16? I suspect it would have been the Harrier (Hawker Siddeley needed an order), and that the Jaguar would not have been procured – it was always the wrong (or certainly over-elaborate) aircraft for an advanced jet trainer (the original intention), and with its relatively high wing loading, was not optimised for CAS. In any event, the UK became involved in Jaguar as part of a collaborative package agreed with the French, where our real interest lay in the Anglo-French Variable Geometry [AFVG] aircraft, which was killed off by [President Charles] de Gaulle in 1967.

'The Lightning was planned to be run-on in the Air Defence/Interceptor role into the late 1970s and, if the TSR2 had survived, I very much doubt that the RAF could have afforded before then a new fighter as well as the TSR2 and the Harrier.

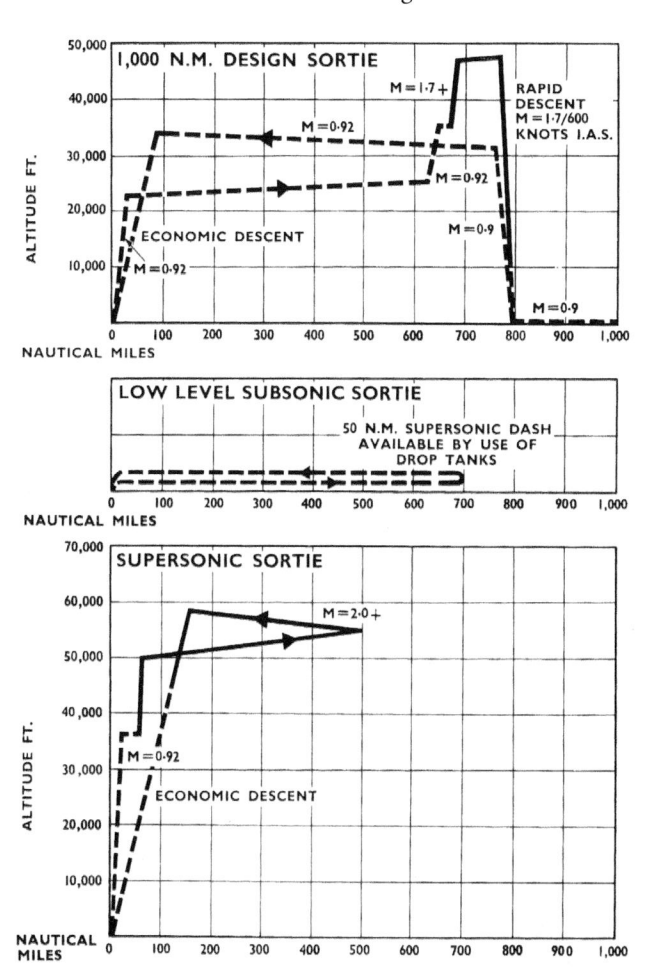

'By the mid-1970s, the need for a highly agile fighter like the F-15 or a multi-role fighter ground-attack aircraft like F-16 or F/A-18 had been widely recognised throughout NATO. There was also the requirement to replace the F-104 and, in France, the Mirage III. Thus an opportunity existed for a collaborative programme in Europe as an alternative to procurement of an American fighter. Industry in the UK would have pushed hard for a European programme for an agile fighter, as would the RAF. But that option was effectively ruled out following the cancellation of the TSR2 and AFVG, and with the Tornado programme launched instead because industrial, economic and political arguments *de facto* forced the RAF down the Tornado ADV [Air Defence Variant] path.

'So, if the TSR2 had survived, it is likely that the UK or Europe would have developed an EFA [European Fighter Aircraft] type ten years earlier than was the case. The lessons learnt during the Jaguar and Tornado collaborative programmes would then have been learnt on the EFA programme instead, but nonetheless a good product would probably have resulted.

The TSR2 subsonic and supersonic mission radii. (Brooklands Museum)

High and proud. (Brooklands Museum)

'Under this plot, there would almost certainly not have been a Buccaneer in service with the RAF. I believe the RAF would have had only the TSR2 and the Harrier in the offensive roles. In the longer term, therefore, the RAF's combat aircraft frontline would have been – TSR2, Harrier and the Lightning replacement. As it was, in 1982, we had the Lightning, Phantom II, Harrier, Jaguar and Buccaneer in service. We also still had some Canberras operating in the recce role up to 2006. We thus had six types instead of three. Moreover, without the TSR2, the V-bombers had to be run on for longer than necessary, awaiting the entry into service of the Tornado GR1.'

One theme running through this saga is that there was no 'Mr TSR2' empowered with keeping the project on time and on budget – nobody in a position to say 'No' to anything. Sir George Edwards of Vickers would have been the man for the job, and as the architect of BAC's survival and recovery after the TSR2 cancellation, he should have the last word:

'The TSR2 was unique inasmuch that it was not only intended as a wholly new and very advanced weapon system for the RAF but was also a politically-charged instrument of the major rationalisation of the British aircraft industry. It was left to [Minister of Defence] Duncan Sandys, who came on the scene with some "golden welding flux" in the form of the TSR2 and other military orders. Vickers was given the contract for the TSR2, provided that EE did half the work, and together we put a lot of effort into it. The really important qualifying factor was that the TSR2 was a very significant part of the BAC workload. Representing about half of that for the military side of the Company, it was expected to have continued as such well into the 1970s. It was a "Weapons System Concept", a new piece of Government book-keeping which meant that the cost of developing its sophisticated equipment and the production investment were all loaded solely onto the bill.

'It was no joke pulling BAC together after the sudden loss of such a large part of the forward workload. The production and development teams at Warton and Preston were both very badly hit. Even when one considers that those two teams under Sir Frederick Page have since played the absolutely dominant part in the design and manufacture of the Jaguar and Tornado, and are the most able and experienced partner in the Eurofighter Typhoon, we had to close a factory, and in the way that these things go, we closed the ex-Hunting plant at Luton. This was a highly-efficient, low-cost, little factory that had nothing to do with the TSR2, but the work had to go North in order to preserve the factories there, and their exceptional military expertise.

'Let's not forget that at the same time as we were grappling with the TSR2 and trying to put together BAC, we were getting the Anglo-French Concorde launched as well. The success of Concorde as an industrial collaboration provided a solid foundation with which to go forward with the Jaguar and Tornado programmes in international collaboration. Together with the Anglo-American Harrier, they have since provided the backbone of the British frontline military aircraft programme and given it a degree of stability that it had not had before – and now a commanding position in Eurofighter Typhoon.

'The extraordinary resolution that we had to summon – simultaneously to overcome the TSR2 cancellation, to cement the BAC organisation and to launch the international dimension – did ensure that the RAF eventually received the right and the best home-grown aircraft, which have certainly proved their worth in live combat on the winning side. This is the most satisfying conclusion that we can all come to, because we must never forget that the RAF is, and always will be, our most important customer.'

FURTHER READING

Burke, Damien, *TSR2 – Britain's Lost Bomber*, Crowood Press, 2010
Lucas, Paul, *TSR2 – Lost Tomorrow of an Eagle*, SAM Publications, 2009
Moore, Richard, *Nuclear Illusion, Nuclear Reality*, Palgrave Macmillan, 2010
RAF Historical Society Journal, *TSR2 with Hindsight*, 1998
Ransom, Stephen and Fairclough, Robert, *English Electric Aircraft and their Predecessors*, Putnam, 1987
Wood, Derek, *Project Cancelled*, Macdonald and Janes, 1975
Wynn, Humphrey, *RAF Nuclear Deterrent Forces*, HMSO, 1994

INDEX

References to illustrations are shown in **bold**, with the caption page in brackets if not on the same page.

acceleration 10, 32, 47, 49
Air Ministry 4, 5, 6, 7, 35, 57, 58, 59, 60, 61, 64, 67, 68
Air Staff 5, 6, 7, 8, 54, 55, 57, 60, 67, 70, 71, 76
    Chief of the Air Staff (CAS) 8, 61, 64, 65, 67; Deputy Chief of the Air Staff (DCAS) 36, 54, 59
airfields 6, 7, 16, 42, 53, 66, 68, 70, **71**
airframes 10, 11, 12, **20**, 31, 37, 57, 58, 67, 69, 71
Akrotiri 53, 67, 72
altitude 11, 12, 17, 35, **42**, 66, 67, 68, 70
    high 12, 23, 44; low 5, 8, 10, 12, 18, 23, 44
automatic flight control system (AFCS) 24, 25, 31, 34, 40, 69
autostabilisation 31, 47, 48
avionics 12, 13, 18, **27**, **32**, 33, 34, 45, 54, 69, 72, **73**, 76
    GEC-Marconi Avionics 24, 34, 37
Avro 6, **25**, 53
azimuth 24, 31, 34

Baghdad Pact/Central Treaty Organisation (CENTO) 5, 60, 67
Beamont, Roland 'Bee' 39, 40, 41, 42, **43**, 44, 45, 50, **51**
Blackburn Buccaneers 6, 12, 24, 30, 31, 32, 35, 53, 58, 61, 70, 71, 73, 78
bombs 12, 16, 22, 23, 32, 33, 34, 52, 53, 66, 67, 68, **69, 70**, 71, **73**
bomb-bays **8**, 11, **61**, **64**, **68**, **69**, **70**, **76**;
    Bomber Command 19, 53; V-bombers 4, 8, 52, 53, 54, 70, 71, 76, 78
Boscombe Down 37, 38, **41**, 42, 43, 51
brakes 15, 39, 40, 46, 47, 48, 49
    airbrakes **13**, 15, 17, 47, **48**, 55
Bristol Siddeley 7, 13, 18, 57
British Aircraft Corporation (BAC) 7, 12, 17, 18, 32, 53, **54**, 55, 57, 65, 78, 79

cameras 27, 71
    Q-band 24, 27, 29
Canberras 4, **5**, 7, 8, 9, 10, 11, 17, 24, **26**, 35, 37, 40, **44**, 54, 57, 58, 59, 61, 67, 71, 72, 73, 76, 78
B(I)8 **5**, 17, 71
climb/ascent 17, 18, 25, **41**, 47, 67
cockpits 13, 14, 23, 25, 27, 34, 35, 36, 40, **45**, **46**, 47
computers 6, 23, 27, 28, 33, 34, 36, 37, 66, 69
    digital 12, 32, 33, 68; navigation 23, 27;
    Verdan 24, **33**, 34, 35, 37, 68
Concorde 19, 20, 21, 31, 32, 46, 79
contractors 6, 7, 33, 37, 57
Controller Aircraft (CA) 8, 44, 61

Decca 24, 34, 37
    Doppler 14, 22, 23, 24, 26, 32, 34, 37
Dell, Jimmy 40, 42, 44, 45, 51, 73
descent 41, 50, 67

Edwards, Sir George 18, 31, 65, 78
EMI 24, 26, 27, 34
engines 7, 11, 13, 14, 15, 16, 17, 18, **19**, **20**, 21, 35, 38, 39, 40, 41, 43, 46, 47, 48, **52**, 57, 59, 69
    Olympus 14, 18, 19, **20**, 21, 39, 69; 22R 18, 19, 21, 38
English Electric (EE) 6, 7, **10**, 11, 12, 13, 17, 18, 28, **44**, 45, 66, 70, 78
    P17 10, 11, 18
Europe 5, 6, 17, 27, 66, 77

Ferranti 22, 24, 27, 28, 29, 31, 34, 37, 69
fighters 44, 47, **51**, 55, 65, 77
final assembly **18**, **25**, **29**, **30**, 38, 43
fins **2**, 11, **13**, 15, 16, 31, 44, 68
forward-looking radar (FLR) 22, 24, 25, 26, 28, 30, 31, 32, 34, 53, 69
fuselage 11, 13, 14, 15, 16, 17, 18, **23**, **25**, **43**, 50, 68, 71
    centre 12, 13, 14; forward **8**, **73**; nose 13, 71; rear 12, **13**, **14**, 15, 18

General Dynamics 14, 35, 53
    F-111 14, 53, 54, 55, 56, 57, 60, 64, 72, 76
General Operational Requirements (GOR) 5, 6
GOR339 5, 6, 7, 8, 10, 11, 12, 57, 58, 59, 61

Harrier 55, 71, 76, 77, 78, 79
Hawker Siddeley 6, 7, 28, 77
head-up display (HUD) 14, 24, 25, 31, 34, 35, 46, 47, 69
hydraulics 13, 15, 32, 39, 55

intakes 14, **19**, 20, 21, 40, 46, 69
interceptors 8, 18, 77

Jaguar 37, 71, 77, 78, 79

landings 12, 16, 34, 39, 40, 41, **43**, 44, 48, 50
lifts 10, 11, 16, 17, 47
Lightning 8, 10, 18, 28, 30, 31, 32, 40, **44**, 45, 47, **51**, 53, 72, 77, 78
F-35 33, 52

Mach 20, **44**
    0.8 42; 0.88 17, 67; 0.9 44, 48, 67, 68; 0.92 67, 68; 1.12 42, 48; 1.6 66; 1.7 7, 35, 68; 2.0 7, 67, 72; 2.20 17, 45, 73; 2.25 17, 45, 66, 73
Management 24, 56, 57, 59, 61
McDonnell F-4 55
    Phantom II 55, 64, 71, 78
missiles 5, 12, 28, 33, 35, 52, 54, 57, 66, 68, 70
Minister of Supply (MoS) 7, 58, 59, 61
Ministry of Aviation (MoA) 17, 18, 29, 35, 53, 55, 58, 59, 60, 61, 64, 68
Ministry of Defence (MoD) 17, 35, 58, 59, 60, 61, 64, 65
moving map 24, 34
    displays 27, 28, **46**

NATO 35, 59, 60, 67, 77
navigators 5, 13, 17, 23, 25, 34, 35, 37, 39, 42, 46, 47, 53, 68, 71, 73
    inertial 14, 22, 24, 26, 33
nosewheels 16, 17, 47
nozzles 15, 18, 20, 46
nuclear weapons 8, 66, **69**

Operation *Granby* **61**, 65, 73
Operational Requirements (OR) 47, 72, 73
    OR343 8, 11, 57

Page, Sir Frederick 11, 31, 65, 66, 79
parachutes **13**, 15, **30**, 39, 40, 45, 48, 67
pilots 5, 13, 14, 17, 24, 25, 29, 35, 40, 42, 44, 45, 46, 47, **53**, 54, 57, 65, 72, 73

autopilot 23, 31; test 31, 35, 39, 45, **51**, 72
prototypes **25**, **29**, 33, 37, 38, 43, **52**, 54, 56
pylons **49**, 66, **71**

radar-absorbing material (RAM) 69, 70
radars 5, 6, **8**, 11, 12, 13, 14, 22, 23, 24, 25, 26, 27, 29, 30, 66, 67, 69, 70
    Doppler 14, 22, 23, 24, 26, 32, 34, 37
radio countermeasures (RCM) 7, 70
radius of action 5, 8, 17, 66, 68
reconnaissance 5, 6, 7, **8**, 11, 23, 24, 26, 27, 35, 37, 52, 58, 67, 70, 71, **73**
refuelling 6, 8, 16, 17, 57, **73**
retraction 15, 47, 50, 55
Royal Aircraft Establishment (RAE) 28, 33, 70
Royal Air Force (RAF) 4, 6, 8, 19, 33, 44, 45, 51, **53**, 57, 58, 60, **61**, 64, 65, 66, 67, 70, 71, 72, 76, 77, 78, 79
    Germany 53, 71; No. 16 Sqn **5**, **56**, 71
runways 6, 16, 38, 48, 50, 51, **61**, 66, 73

Short Take-Off and Landing (STOL) 4, 16, 55, 76, 77
HS681 4, 55
sideways-looking radar (SLR) 6, 23, 26, 27, **28**, 34, 36, **46**, 69, 71
simulators **36**, 40, 45
sorties 7, 36, 41, 44, 48, 68, 71
Southeast Asia Treaty Organisation (SEATO) 5, 60, 67
supersonic 6, 12, 20, 21, 35, 42, 44, 55, 67, **77**
speeds 6, 11, 12, 18, 67

tactical strike 5, 7, 8, 58, 66, 70
tailplanes 11, **13**, 15, 16, 47, 48
takeoff 6, 7, 11, 12, 16, 17, 18, 39, 40, 47, 57, 58, 66, 68, 73
TFX 53, 54, 57
Thorneycroft, Peter 39, 52, 58, 64
throttles 15, 19, 41, 46
thrust 6, 7, 11, 12, 14, 17, 18, 19, 20, 35
Tornado 17, 32, 51, 69, 70, 71, 72, 73, 76, 77, 78, 79
GR1 17, 69, 76, 78
TSR2 4, 7, **8**, 11, 12, 13, **14**, **15**, 16, 17, **18**, 19, **20**, 21, **22**, **23**, 24, **25**, **26**, **27**, **28**, **29**, **30**, 31, **32**, **33**, 34, 35, **36**, 37, 38, 40, 41, 42, **43**, **44**, **45**, **46**, 49, 50, **51**, 52, **53**, **54**, 55, **56**, 57, 59, 60, **61**, 64, **65**, 66, 67, **68**, **69**, **70**, **71**, 72, **73**, **76**, **77**, 78, 79
XR219 **29**, **38**, **39**, **40**, **41**, **42**, 43, **44**, 47, **48**, 54, 56, **58**, **60**, **64**, 66

undercarriage 11, 12, **13**, 14, 15, 16, 38, 40, 41, **42**, **43**, 44, 47, **49**, **50**, 51
US Air Force (USAF) **5**, 44, 53, 72
F-105 **5**, 72

VC10 airliners 18, 32, 34
velocity 24, 32, 33
Vickers 6, 7, 12, 13, 18, 37, 65, 66, 78
Vickers-Armstrongs 7, 12, 24

Warton 10, 28, 38, 42, 45, **51**, **55**, **65**, **66**, 79
Weybridge 7, 12, **14**, **15**, 24, **29**, 38, 54, 56, 71
Whitehall 59, 61, 65
Wilson, Harold 55, 56, 60, 64
wings **10**, 11, 12, **13**, **15**, 16, 18, 19, 35, **49**, 53, 67, 71, 77
wingtips 16, **29**, 45, **49**

X-band 24, 26, 29, 69, 70